GM TURBO 400 TRANSMISSIONS

Cliff Ruggles

CarTech®

CarTech ®

CarTech®, Inc.
6118 Main Street
North Branch, MN 55056
Phone: 651-277-1200 or 800-551-4754
Fax: 651-277-1203
www.cartechbooks.com

Edit by Scott Parkhurst
Layout by Monica Seiberlich

ISBN 978-1-934709-20-7
Item No. SA186

Library of Congress Cataloging-in-Publication Data

Ruggles, Cliff.
  How to rebuild & modify GM turbo 400 transmissions / Cliff Ruggles.
    p. cm.
  ISBN 978-1-934709-20-7
  1. General Motors automobiles—Transmission devices, Automatic.
2. General Motors automobiles—Transmission devices, Automatic—Maintenance and repair. I. Title.

TL263.R845 2011
629.2'4460288—dc22

2010050848

Written, edited, and designed in the U.S.A.
Printed in China
10

Title Page:

Get a firm hold on the input shaft and pull the forward clutch drum from the case.

Back Cover Photos

Top Left:

Lower the oil pump into place, and install the bolts with new seals from the rebuild kit.

Top Right:

The entire output assembly can now be lowered into the case. Do not dislodge the low band during this procedure. Once it is fully into position, you can install the sun gear in the center of the second picture. Note the orientation of the drum with the low band. Also note that the sun gear is installed with the inner chamfered edge down toward the back of the case. The flat side faces up toward the center support and upper Torrington bearing.

Middle Left:

Install a new thrust washer on the lower end of the output shaft assembly, and install the speedometer drive gear if removed previously.

Middle Right:

With a flat-tip screwdriver or other suitable tool, remove the snap ring to separate the lower planetary assembly from the output shaft assembly. This allows access to the two sets of Torrington bearings, and to replace the output shaft upper bushing.

Bottom:

Our transmission is in place, strapped down, and ready to safely install in the vehicle.

OVERSEAS DISTRIBUTION BY:

Europe
PGUK
63 Hatton Garden
London EC1N 8LE, England
Phone: 020 7061 1980 • Fax: 020 7242 3725
www.pguk.co.uk

Australia
Renniks Publications Ltd.
3/37-39 Green Street
Banksmeadow, NSW 2109, Australia
Phone: 2 9695 7055 • Fax: 2 9695 7355
www.renniks.com

Canada
Login Canada
300 Saulteaux Crescent
Winnipeg, MB, R3J 3T2 Canada
Phone: 800 665 1148 • Fax: 800 665 0103
www.lb.ca

CONTENTS

ACKNOWLEDGMENTS

Thank you, again, to Kris Abrahamson, of Continental Torque Converters, Inc., for his knowledge on torque converters. I also thank Transtar Industries, Inc., for their wealth of knowledge on improving shift function, and TCI Automotive, Inc., for their excellent TH400 Transbrake.

I am grateful to my friends, family, and especially my wife, Deb, for being there for a whopping 30 years as of the time of this writing!

Finally, to my longtime friend Ray Klemm: Thanks for always being there, and for the countless hours that you contributed to this project.

PREFACE

I have been a dedicated muscle car enthusiast for the better part of my adult life. With no regret whatsoever, I can say that there are few things more enjoyable than being able to work on your own car or truck, and get perfect results from your efforts. To this day, I have never understood why hobbyists farm out many of the components of their cars to others, instead of rebuilding them in their own shops or garages. Transmissions are at the top of the list for the items that get farmed out. There has always been a mystique about them that makes even the most skilled automotive technicians avoid them. There are also the pretty hefty bills that can come from not doing them at home.

Similar to most other things associated with this hobby, you can get results that are just as good, or better, and save money by doing it yourself. The key component here is information, followed closely by motivation. So this book is for the hobbyist—to help provide information and motivation.

My need for information emerged dramatically on a Saturday afternoon in the fall of 1978. I had decided to tackle a TH400 transmission rebuild—in the small garage next to our house—and without any assistance. Not that I really wanted to do it; at that time I would have rather been squirrel hunting or going someplace with my friends. I still chuckle every time I lift the valve body off a transmission. I think back to the six check balls that were rolling across the concrete floor when I had removed the valve body and turned the unit over to allow the fluid to drain out. At that moment I was really thinking I should have carried it to a shop. But my income level at that time just didn't permit a high-performance muscle car that wasn't moving because I had fried its transmission. I managed to find five out of the six of the check balls, and after careful observation could see the tracks in the case where they were located. The story had a good ending, although that particular rebuild took me several weeks, and I wasn't confident the car was going to move under its own power when I lowered it to the ground for a test drive!

More than three decades and hundreds of units later, I'm equipped with quite a bit more knowledge, and a tool box full of homemade tools dedicated strictly to transmission work. I don't give a second thought to stripping any automatic transmission completely down to a bare case, and then putting it back together with my large assortment of hand-crafted custom tools, and self-taught special procedures.

I set two goals for this book: to provide others with the information I've learned over many years and to present the information in a way that makes this task well within reach of the average enthusiast. I also want to help with motivation: You really can do the work at home, by yourself, and can make a lot of what you need to accomplish the task from common hand tools or other items you are likely to find in your own garage.

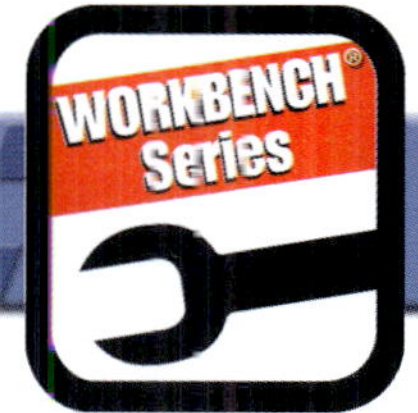

WHAT IS A *WORKBENCH*® BOOK?

This Workbench® Series book is the only book of its kind on the market. No other book offers the same combination of detailed hands-on information and revealing color photographs to illustrate transmission rebuilding. Rest assured, you have purchased an indispensable companion that will expertly guide you, one step at a time, through each important stage of the rebuilding process. This book is packed with real world techniques and practical tips for expertly performing rebuild procedures, not vague instructions or unnecessary processes. At-home mechanics or enthusiast builders strive for professional results, and the instruction in our Workbench® Series books help you realize pro-caliber results. Hundreds of photos guide you through the entire process from start to finish, with informative captions containing comprehensive instructions for every step of the process.

The step-by-step photo procedures also contain many additional photos that show how to install high-performance components, modify stock components for special applications, or even call attention to assembly steps that are critical to proper operation or safety. These are labeled with unique icons. These symbols represent an idea, and photos marked with the icons contain important, specialized information.

Here are some of the icons found in Workbench® books:

Important!—Calls special attention to a step or procedure, so that the procedure is correctly performed. This prevents damage to a vehicle, system, or component.

Save Money—Illustrates a method or alternate method of performing a rebuild step that will save money but still give acceptable results.

Torque Fasteners—Illustrates a fastener that must be properly tightened with a torque wrench at this point in the rebuild. The torque specs are usually provided in the step.

Special Tool—Illustrates the use of a special tool that may be required or can make the job easier (caption with photo explains further).

Performance Tip—Indicates a procedure or modification that can improve performance. Step most often applies to high-performance or racing engines.

Critical Inspection—Indicates that a component must be inspected to ensure proper operation of the engine.

Precision Measurement—Illustrates a precision measurement or adjustment that is required at this point in the rebuild.

Professional Mechanic Tip—Illustrates a step in the rebuild that non-professionals may not know. It may illustrate a shortcut, or a trick to improve reliability, prevent component damage, etc.

Documentation Required—Illustrates a point in the rebuild where the reader should write down a particular measurement, size, part number, etc. for later reference or photograph a part, area or system of the vehicle for future reference.

Tech Tip—Tech Tips provide brief coverage of important subject matter that doesn't naturally fall into the text or step-by-step procedures of a chapter. Tech Tips contain valuable hints, important info, or outstanding products that professionals have discovered after years of work. These will add to your understanding of the process, and help you get the most power, economy, and reliability from your engine.

TOOLS AND EQUIPMENT

Completely and correctly rebuilding an automatic transmission is not beyond the capabilities of the average automotive enthusiast. But there is something about it that keeps away even the most skilled automotive technicians. I always hear the same thing: "I don't have all the special tools to rebuild them." So the tranny is usually carried off to a specialty shop instead.

It is true that a few tools, which may be called "special," are required for certain operations I cover in this book. But the good news is that the majority of the tools required are simply ordinary hand tools. Even better news is that, with a little extra effort, you can improvise or fabricate a tool for most any particular operation.

Safety Equipment

Safety equipment is first and foremost on my list. Automatic transmissions use pressurized fluid. During a rebuild, you come into contact with transmission fluid and cleaning solvents. You also use compressed air for drying parts, and for air-pressure-testing clutch packs, servos, etc.

Eye protection is a must anytime compressed air is used. When cleaning parts, an injury can result easily and very painfully if high-detergent transmission fluid or a piece of debris is blown into your eyes.

Nitrile gloves are good for keeping solvents and transmission fluid off your hands. They also provide some protection from getting cut on the jagged edges inside the cases. The quality of the gloves currently available varies considerably. The better-quality gloves are thicker and more resistant to solvents such as

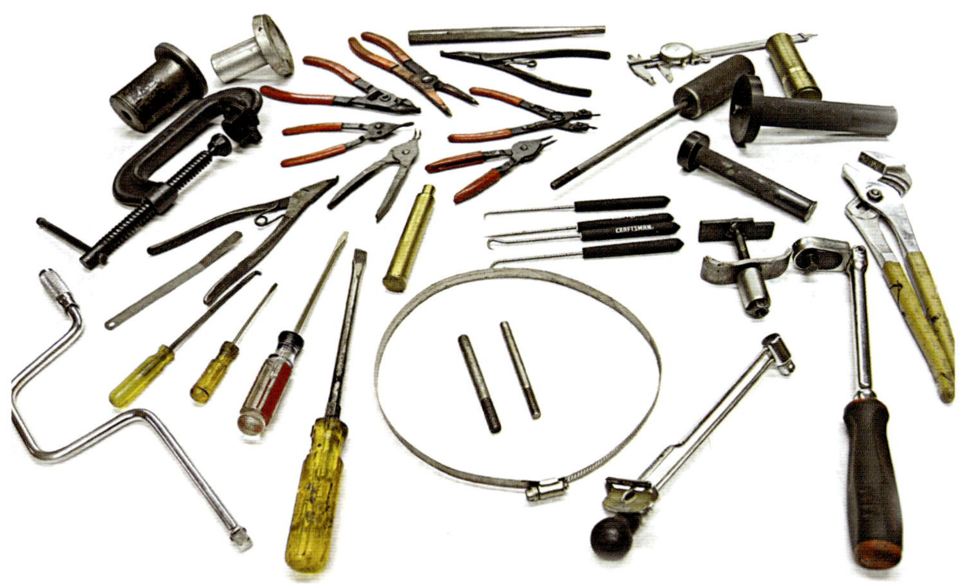

Most tools that are needed to rebuild your transmission are those that are found in most tool boxes. Those tools can be modified to fit the task at hand or even fabricated from common items, such as putting two large hose clamps together to make an oil pump alignment tool.

brake cleaner, which works well to remove oily residue from parts before assembly.

Heavy-duty chemical-resistant gloves also provide protection. Wear them when cleaning the transmission case with heavy-duty degreasers or other harsh chemicals.

Hearing protection is also essential. During the rebuild, you may use compressed air to clean parts and dry them. When high-pressure air is blown into the case passages, for example, it can create a high-pitched sound that is injurious to your hearing.

Sockets and Wrenches

General Motors used both metric and Society of Automotive Engineers (SAE) fasteners through the years of production. The later units used metric fasteners, but some units have SAE and metric on the same unit. A complete socket set, with 1/4-inch-drive, 3/8-inch-drive, and 1/2-inch-drive sockets are needed.

Make sure you have extensions for each drive. An air ratchet or an impact gun speeds up the rebuilding process considerably, but never use them to tighten fasteners because most of them are threaded into the aluminum case. The aluminum is soft enough that the threads may strip out if you use excessive torque. I prefer a spin-handle wrench for quickly installing bolts and tightening them down. They do not require compressed air and are much less likely to overtighten any fasteners.

A torque wrench is a valuable tool because each fastener size has a torque value based on its diameter and hardness. Still, using one does not make up for common sense, and the long handles on a torque wrench make it easier to overtighten a bolt and strip the threads out of the case. This is a greater concern with the smaller fasteners used to hold the valve body in place.

A good pair of safety glasses with plastic or impact-resistant lenses is a must for transmission work.

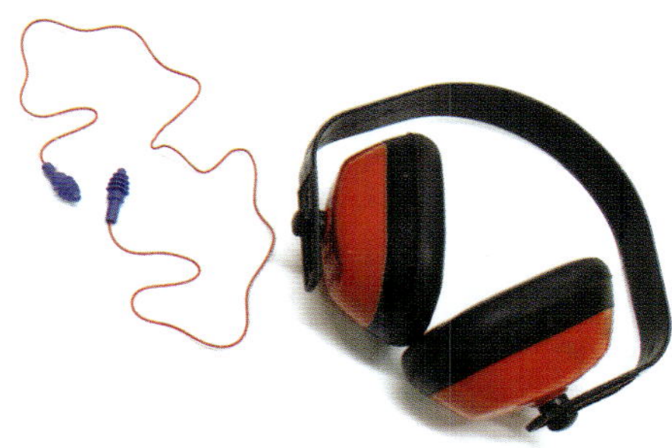

Hearing protection is a must when hammering bushings in place and blowing off parts with compressed air. The tiny passages found in transmissions can produce loud noises when hit with high-pressure compressed air.

Chemical-resistant gloves help protect your hands from cleaning solvents, automatic transmission fluid, and assembly lubricant.

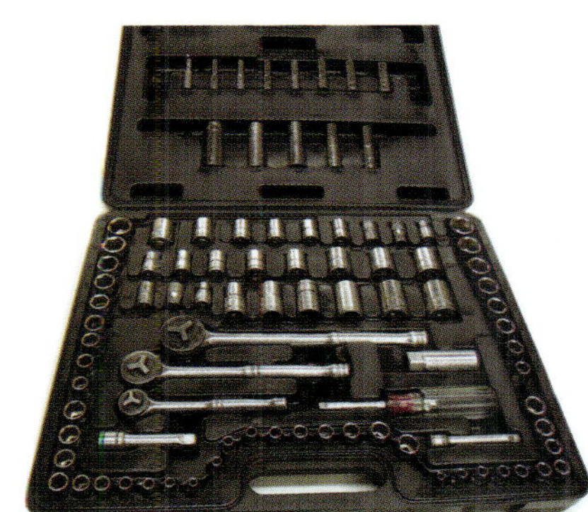

You need complete metric and SAE socket sets. Most fasteners are SAE, but there can be a few metric bolts on later-model units.

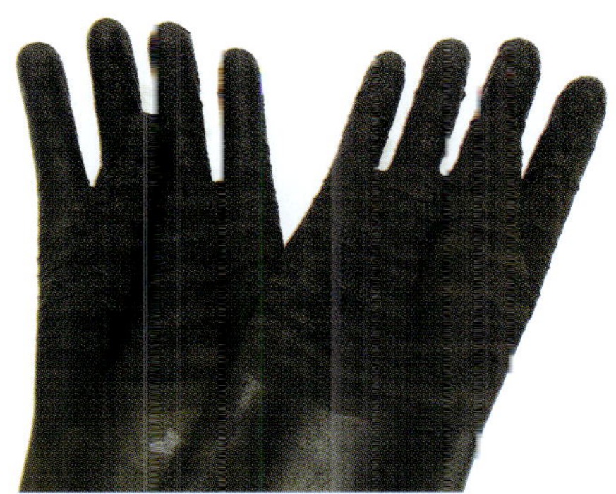

Wear heavy-duty gloves if using a parts washer and solvents. They also offer some degree of protection when handling transmission cases during the cleaning process; sharp edges can leave some pretty nasty cuts.

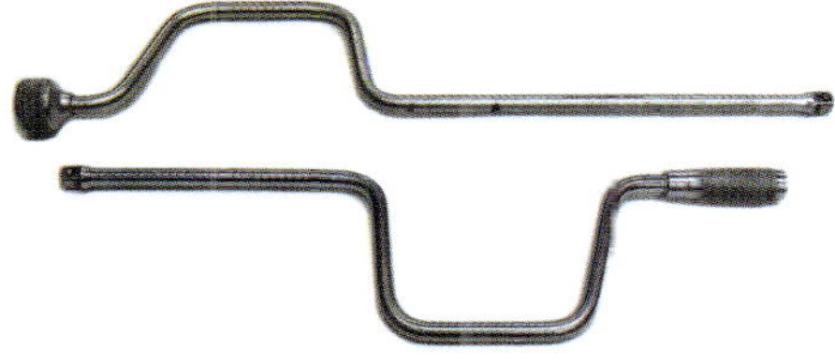

Using a spin-handle wrench speeds the rebuilding process dramatically. It also eliminates the possibility of stripping out any threaded holes in the case when using air impacts or air ratchets.

In addition to a complete socket set, you also need some hand wrenches. I especially like the new styles that have the ratcheting feature on the boxed end. They shorten the time typically required to tighten or remove a fastener that you cannot access with a socket.

Flare-nut or tubing wrenches are made specifically for accessing the nuts that hold steel lines in place. They have additional material and increased integrity to loosen stubborn flare nuts that are used to hold cooling lines to the case. They allow the wrench to slip over the tube and still get a good hold of the flare nut to remove it.

Snap Ring Pliers

You need a variety of snap ring pliers. Some of them must be modified slightly for a specific purpose.

You can also buy pliers with removable tips, so one pair can cover a broad range of applications. Some snap ring pliers are also designed to be convertible. They can be modified in seconds to remove either the inner or outer snap rings.

Flare-nut wrenches are a must for removing and installing cooling lines. They provide access to the nuts but are designed to have increased contact area to help remove and install the nuts without rounding them off.

You can modify a large set of ring expanders. They work well when removing the spring cage retainers on the drums.

You need a variety of snap ring pliers for transmission work. Some are available with removable ends to work with a wide variety of snap rings using the same tool.

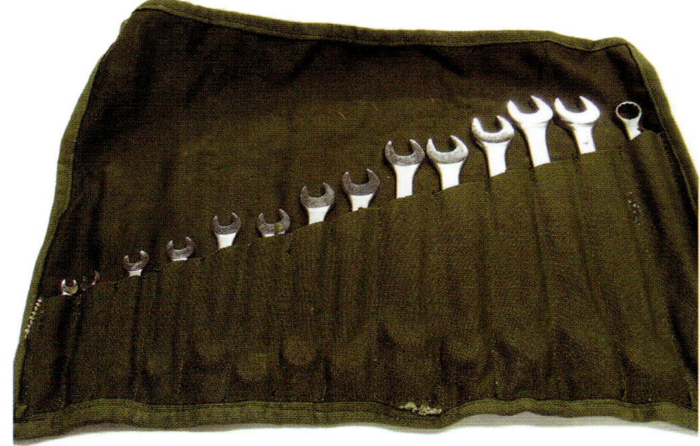

You need standard and metric wrenches. Most fasteners are SAE, but later units incorporated some metric fasteners in a few places.

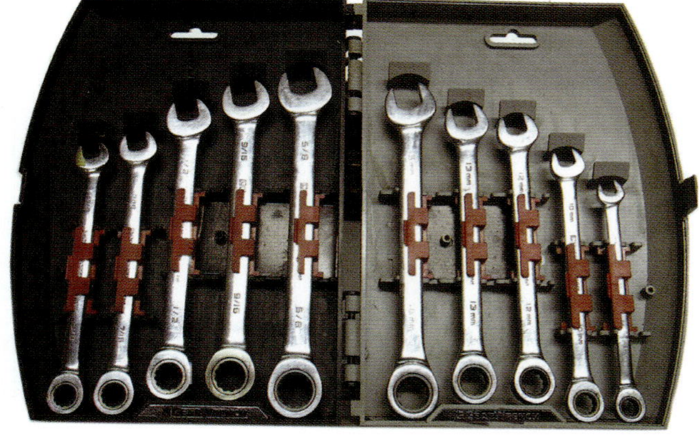

Modern technology has produced some very nice tools, and these wrenches are one of them. They can save a lot of time loosening and tightening fasteners that are not accessible with a socket.

Screwdrivers

A variety of screwdrivers come in handy during a rebuild. Large flat-tip screwdrivers can be used to remove the snap rings in transmission cases, and also to put them back in position. Phillips screwdrivers also double as alignment tools; for example, when putting the pump back on the case. In a couple of minutes with a torch, you can take a couple of small flat-tip screwdrivers and turn them into "friction and steel" removal tools. By bending the ends over, you can use them to reach inside clutch drums or into the transmission case to pull steels and friction plates out. An awl also comes in handy for transmission rebuilding. You can use it to help line up bolt holes with gaskets and separator plates when installing the valve body, and to remove the snap ring that retains the 1-2 accumulator in the case.

Measuring Devices

Transmission rebuilding involves precision and attention to detail. Quite a few items within the assembly need to be measured for specification. A dial indicator is a valuable precision tool for transmission builds. It can be used to check input- or output-shaft endplay before and after the unit is assembled. Since you may have selective shims available to set the shaft endplay, it's preferable to be able to measure all components accurately. It can also be used to measure the inside and outside diameter of bushings and other components. The other end also has a depth gauge. This handy part of the tool indicates how far down inside a drum a particular bushing was installed prior to driving it out for replacement.

A standard zero to 1-inch micrometer works equally as well for measuring frictions (friction plates) and steel plates, as well as the total thickness of a clutch pack.

Large flat-tip screwdrivers help you remove the large snap rings from the cases on the TH400. They can also be used to remove the manual shaft seals from the case.

Phillips screwdrivers make great alignment tools for installing the oil pump, and valve bodies to the case.

You can modify a pair of screwdrivers by heating them and bending the ends over. They have a variety of uses, including pulling steels and frictions out of clutch drums and cases.

Use a dial indicator to determine input shaft endplay. Measure endplay, and then use shims or selective washers to keep it within specifications.

Dial calipers are for measuring thickness or size. They also can be used to measure depth. A good pair is pretty much a mandatory item for transmission rebuilding because you need to measure the thickness of friction and steel plates when stacking clutch drums.

Micrometers are another necessary measuring device. Ball micrometers are used to measure round surfaces and can quickly determine bushing thickness. Flat-tip micrometers measure the diameter of a valve in the valve body.

Spring Compressors

Turbo-Hydramatic 350s require a piston spring compressor inside the case. Many years ago I made a spring compressor, in about five minutes, from two pieces of flat steel bar. It has served me well for decades and the total investment was only a few dollars.

For the Turbo-Hydramatic 400 (TH400), you can fabricate similar tools from steel flat plate and longer pieces of threaded rod. You can use these to compress spring cages on other components, provided they have a hole through the center of them. For drums that do not, hold the component stationary and use some sort of spring compressor to push down on the spring cage to the snap ring.

You need a larger bench-mounted spring compressor for building clutch drums. A shop press works equally as well, although they tend to be much more difficult to set up and a lot slower to use. You can fabricate a spring compressor from a few pieces of angle and square tubing. They are also available through most commercial tool supply stores. The unit I use was built from a few pieces of metal left over from other projects. It mounts easily in a large shop vise when I use it for clutch drums with a protruding shaft. You can drill a hole in your workbench to accommodate the protruding shafts of these drums if you need to, which allows the tool to be used on a flat work surface if a large vise is not available.

Case Holding Fixtures

A transmission holding fixture, if available, makes a build go much easier. One is certainly not required, but being able to turn the transmission into any position and lock it in place helps with the disassembly and reassembly processes. Several different types are available. Conveniently, General Motors cast bosses into the case to accommodate a special holding fixture. Two round pins engage the case, one on each side, and then a long threaded bolt is tightened against the top of the case. Since no oil pan bolt holes are used, the entire transmission can be taken apart and reassembled while in the holding fixture.

Most universal holding fixtures use a couple of bolts into the pan rails. They still provide stability for the unit, and the ability to lock it into any position, but the tool must be removed to install the oil pan and complete the rebuild.

Work Surface

If you are using a holding fixture of any type, it needs to be securely attached to a heavy-duty table or workbench. You may need to counterbalance the workbench to offset the weight of the transmission. You can construct a simple heavy-duty workbench from a couple of 4 x 8 sheets of plywood, 4 x 4s, several 2 x 6s, and some good wood screws. The benches I use in my shop have storage space underneath them. The weight of the parts on the lower shelf provides enough stability so the weight of the transmission doesn't cause the table to fall over. The 8-foot work surface provides plenty of room to lay out all of the parts removed from the transmission.

A spring compressor is a required tool for transmission building. Shown here is a homemade unit fabricated from square stock and angle iron. This allows you to gently compress spring cages just enough to remove and install the retaining rings without damaging the spring cage.

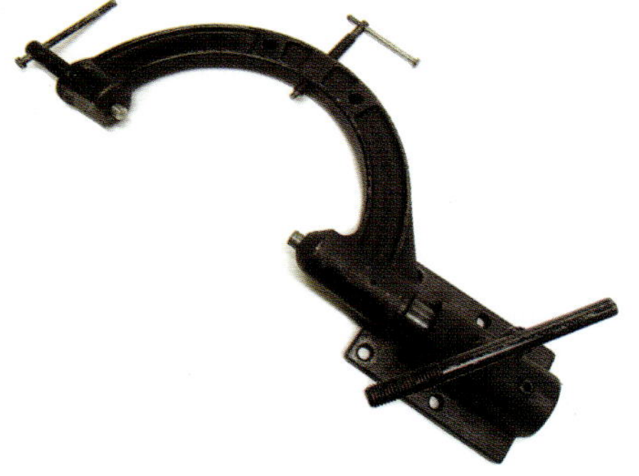

Several types of holding fixtures are available for transmissions. General Motors cast bosses on all TH400 cases to accommodate the appropriate holding fixture.

Universal holding fixtures utilize the pan bolt holes to attach it to the case. The downside of these fixtures is that they must be removed to install the transmission's oil pan.

A suitable working surface is a must for building transmissions. You need a heavy-duty table or workbench to effectively support the weight of the transmission if a holding fixture is used. The more surface area on the top of the bench the better; it allows you to keep the parts separated and organized during the procedure.

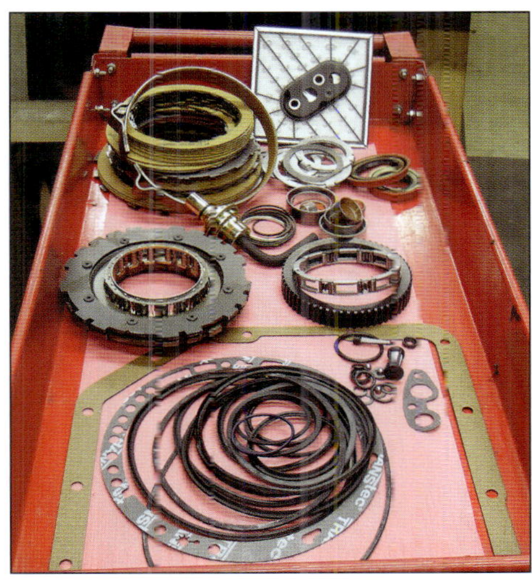

Rolling carts also make a great place to lay out rebuild kits, bushing sets, thrust washers, and other items used to rebuild transmissions. You can pull them right up to the work area without mixing up the new parts with used ones, and push them away when cleaning off and blowing parts dry with compressed air.

Large commercial baking pans keep parts on the bench from rolling onto the floor. They also hold any fluid that drains from the parts when they are removed.

Pans for Small Parts

Large commercial-grade cookie sheets make a great place to put the internals of the transmission as they are removed from the case. The cookie sheets hold the oil draining from the parts, and keep any small parts (such as check balls) from rolling onto the floor. Frictions and bands need to be soaked in clean automatic transmission fluid (ATF) for at least 15 minutes prior to assembly. You can use a cleaned 1.5-gallon ice cream pail, with about a quart of clear ATF, to soak frictions and bands. It has a lid to keep out dirt and debris. It also works well for

You need a container to soak friction and band material prior to installation. Ice cream buckets work well, and they have a lid to keep dirt and debris off the new parts.

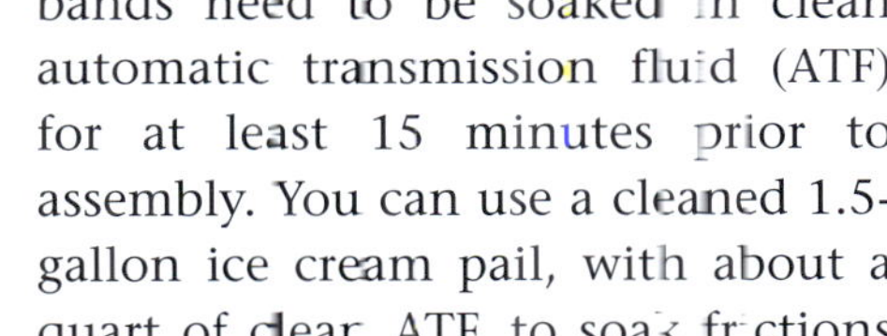

A power washer makes a great transmission case cleaner. Using a "zero" tip, a good power washer blasts off dirt and grease and makes the cleanup process a lot easier.

A gasket scraper, small screwdriver, and wire brushes also help with case cleaning. They help remove really stubborn deposits, such as undercoating and road tar, that the transmission has accumulated over years of service.

Several very good degreasers are available to clean up transmission cases and tail housings that may be covered with dirt and grease. Soak the transmission for several hours and then blast the case with a power washer, or take it to a local car wash.

lubricating Torrington bearings, and dipping apply pistons prior to installing them in the drums.

Cleaning Cases

After the transmission is completely stripped down, the case needs to be cleaned. They are often coated with mud, dirt, grease, undercoating, and road tar. You can make a quick trip to a local carwash and easily remove most of the heavy debris. Spend some additional time with a screwdriver and small wire brush to loosen the stubborn dirt and grease. You may need some solvent to get the tar and undercoating off the case. Brake cleaner is a good solvent to help get oil and grease off the case, and it helps dry it as well. It leaves no oily residue behind, and also works quite well for cleaning internal components.

Several companies make heavy-duty degreasers specifically designed for this purpose. Apply them before power washing or hand cleaning to help loosen thick greasy deposits from the case. Some of these products produce fumes so use them in a well-ventilated area. Also wear a good pair of thick chemical-resistant gloves.

Bushing Removers and Installers

Automatic transmissions have quite a few bushings. Some are relatively easy to access, remove, and install. Others may be quite difficult to get to. If a bushing is readily accessible and flush with the top surface

Brake cleaner helps degrease and dry the case prior to assembly. It doesn't leave behind any oil or other chemicals that could prevent paint from sticking.

Bushing drivers can be fabricated from just about anything, such as old engine valves, or anything with a wide flat surface and some sort of handle.

The lower case bushings for the TH400 should be driven in from the front. You need an extension to gain access to them with a suitable bushing driver.

Complete bushing driver sets, as shown here, have the needed drivers to set any bushings that need to go below flush to a specific depth.

A good bushing driver set is expensive, and really not needed if you are only doing one or two units. Most bushings are driven in flush, so any suitable driver with a broad flat surface and some sort of handle works. For bushings that are driven in below flush, an old bushing can be split and used in combination with a flat driver.

A long tapered punch is a must for bushing removal. Ground it on an angle. The sharp edge catches the bushings and helps remove them from the bore. Even bushings in blind holes can be turned sideways and pried out using a tapered punch.

of the hole, removal and installation is relatively easy. Just about any suitable flat driver can be used to flush mount a bushing.

Some bushings are driven below flush or deep enough to be below a lube oil supply hole, for example. You must record your accurate measurements and carefully install these bushings to the same depth that they were originally at. This requires a bushing driver slightly smaller than

the hole the bushing is driven into.

The lower case bushings may require a long extension to drive the new bushing properly into place. You need a large piece of solid steel pipe to help drive them in.

If a bushing is located in a blind hole, it may be difficult to remove. You can fabricate a tool with a sharp punch. Grind it on a slight angle so it can catch the edge of the bushing and drive it down. You then turn it

sideways in the bore and easily pry it out with a large screwdriver.

Heli-Coils and Bolt Extractors

The TH400 is made of a cast-aluminum alloy. Steel fasteners are used to attach the various parts to them, such as the vacuum modulator, oil pan, etc. It is common to strip the threads out of the case in one or more places. This usually occurs with

It's almost certain that one or several bolt holes for the oil pan require a HeliCoil or thread insert. They are easy to install, and the end result is stronger and holds better than the original material.

You need high-pressure air to blow parts dry after cleaning. It can also be used to pressure test the clutch pack operation. By applying air to the fluid supply holes, you can pressurize the passages with air and verify apply piston movement and whether the seals leak.

An angle grinder equipped with abrasive disks is one of the best inventions of modern times for cleaning stubborn gaskets from machined surfaces. The disks are available in several different materials so they do not damage the soft aluminum surfaces.

the oil pan attaching bolts. In some cases you can use a longer bolt to reach new threads. Another option is to move up to a bolt that is larger in diameter. The best repair is to install a thread insert into the case. This is accomplished by using a special tap supplied with the thread insert kit, then installing a steel threaded insert. This repair is stronger than the original material. Several companies make threaded insert kits. They come with a tap, several threaded inserts, and a special tool to install them.

Air Tools

Air tools are helpful in transmission rebuilding. They can considerably speed up the time needed to install and remove fasteners. High-pressure compressed air is also needed to bench test clutch packs and other components during the rebuilding process.

You need a blowgun with a long tip to access small holes in the case, the oil pump, and the clutch drums, in order to air test them. You can use a rubber tip or shop rag to help seal off the air to listen for any leakage at the newly installed seals. The blowgun can also be used to dry off parts and blow dirt off the case during cleaning.

Having compressed air is also a big advantage when cleaning gasket surfaces. You can use an air-driven disc grinder to remove stubborn gasket material and gasket sealers. The disks are available in several varieties, and the finer ones remove gasket material without damaging the aluminum surface where the gasket seals.

You can use impact guns to remove stubborn fasteners, such as those that attach the tail housing or rear transmission mount to the case. You can use a smaller, 3/8-inch impact gun to remove oil pan and

valve body attaching bolts. An air ratchet comes in handy for these tasks as well, but is not as fast as an impact gun. However, if you use an impact gun to tighten fasteners, it should be set low enough to just snug them up. Apply any final torque to them with a torque wrench to minimize the possibility of stripping the threads out of the case.

Punches, Chisels and Files

A variety of punches and chisels are required for transmission work. They are used to "stake" parts in place, and to remove and install roll pins. Many bushings are difficult to remove without splitting them first, especially if they are in blind holes. You can grind a round punch at a slight angle and use it to catch the edge of the bushing. This drives it down on one side and facilitates removing it from the hole. Before removing any bushings, measure the

Air tools can save a lot of time during a transmission build. Even so, make sure they are going in the right direction when removing bolts, which are easy to strip out of the soft aluminum cases.

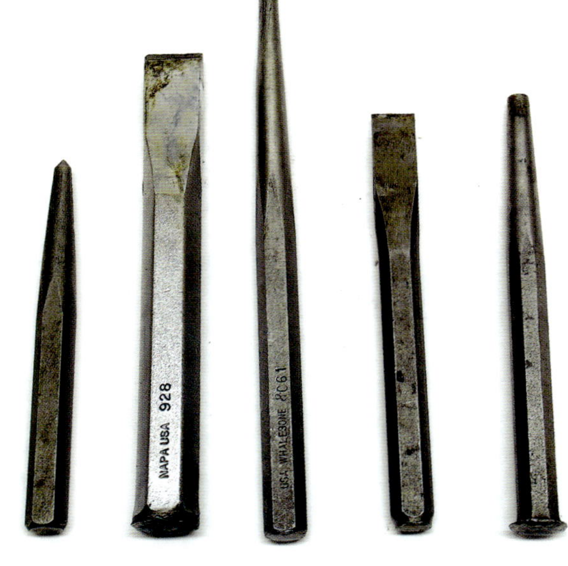

Punches and chisels are handy items for transmission building. There are roll pins that need to be removed and installed, and you may have to split a bushing to get it loose from its bore.

Jeweler's files are used to remove sharp edges. In almost all cases, the manual shaft needs some work to get it out of the case to access the manual shaft seal. The nut on the end usually raises the metal enough so it does not pull through the case. It also leaves a knife-like edge that needs to be removed or it will cut the new seal when it is installed.

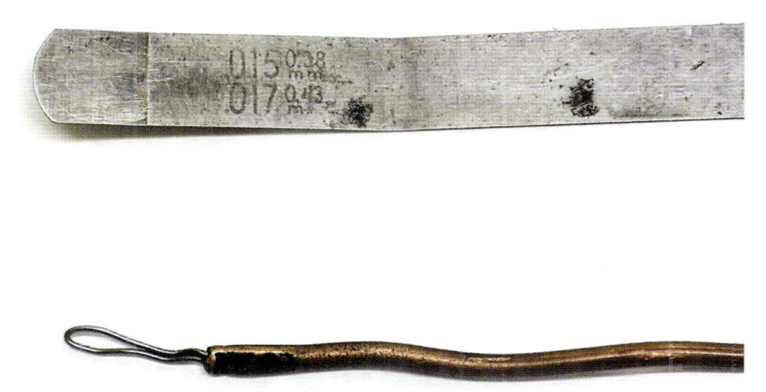

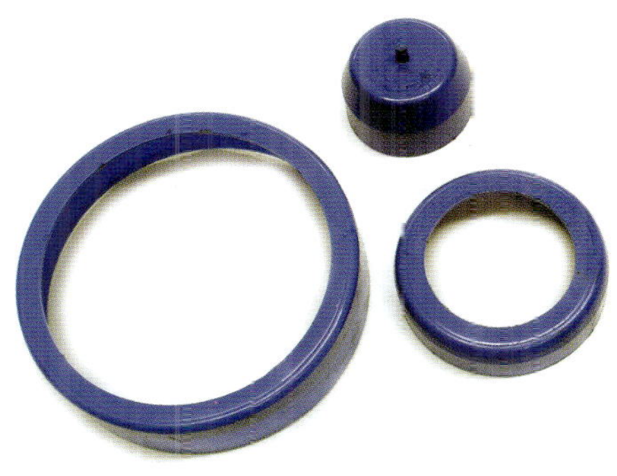

The factory provided chamfered edges to help with seal installation. You still need to help the seals get past the edges without damaging them. A piece of small-diameter wire, crimped in a piece of tubing, makes a great lip seal installation tool. A feeler with a bend on the lower portion works equally well.

Plastic lip installation seals are available, but really not needed with the TH400. They are still a good idea if within the budget, or if you plan on building multiple units.

depth they are set at so you can correctly install the new bushing.

You may need files to remove material from the manual shafts so they can be removed from the case. Most manual shafts have a lip that prevents them from sliding out of the case when you need to access the manual shaft seal. A few seconds of work with a small jeweler's file removes any excess material and they slide right out of the case.

Lip Seal Installation Tools

Lip seals are used to apply pistons and servos to keep hydraulic fluid and pressure behind the piston. The lip on the seal is pushed tight against the bore it rides in, creating a positive seal. Lip seals can be somewhat difficult to install; there is usually an inner and outer seal on each piston.

Special seal installation tools are available for all transmission models. They keep the lip seals from getting torn during the installation process.

There is a slight factory-machined chamfer on the edge of each drum, to help with seal installation. Rather than buy special seal installation tools, you can use a feeler gauge or make your own seal installation tool from a small piece of copper or steel tubing and smooth wire. Insert a loop of wire into the open end of the tubing and crimp it tightly. Use the loop of wire to get the lip of the seal being installed past the chamfered edge without tearing it.

It takes some practice and patience to install lip-type seals without damaging them. The key here is to never force them into place or try to push the seal past a sharp lip without using some sort of installation tool. Air checking them, after the drum is assembled, ensures that you have not ripped or damaged the seals.

Friction Alignment Tool

The intermediate clutches in the TH400 transmission are located in the case. The back of the direct drum is splined to accommodate the intermediate friction plates. In order to install the direct drum, you must carefully align the friction in the center of the case. This allows the direct drum to drop fully into position. An alignment tool can be quickly fabricated by using an old outer race and welding a handle to it.

Another method that can be used here is to blow compressed air

The intermediate frictions must be in perfect alignment to install the direct drum into the TH400 case. A quick and easy way to align them is to use an outer sprag race. I have taken on old race and welded a handle to it to make it easier to use. Alternatively, you can apply compressed air to the intermediate clutch pack through the center hole in the case. You can apply the air to "bounce" the clutch pack and help align the frictions while the drum is being installed.

A couple of 5/16-inch bolts can be fabricated and used as alignment studs to install the oil pump.

into the oil-supply passage for the intermediate clutch pack. This "bounces" the frictions and steels. Turning the direct drum while it is being installed and bouncing the intermediate clutch pack with compressed air aligns the frictions with the splines on the direct drum. It takes some patience when using this method, but it is an effective substitute for making an alignment tool for the clutch pack.

Oil Pump Band

Oil pumps are made in two halves. You must take them apart to access the pump gears, bushing, and seal. When they are ready to assembled, the two halves must be aligned or the pump will not fit back into the case. You can make a band from a large single hose clamp or by putting several large hose clamps together. Leave the pump bolts finger tight while tightening the band. Use a Phillips screwdriver or awl to align the bolt holes before tightening the pump bolts. It helps to drop the

You can use a large hose clamp to align the pump half when tightening the pump bolts. If not in alignment, the pump is difficult to install in the case and the bolts may not line up enough to get them started.

pump attaching bolts through the holes while the pump bolts are tightened. This ensures that the halves do not slip and get out of alignment. After the bolts are tightened, you should test fit the pump into the case and start all the pump attachment bolts by hand. Do not install any seals for the test fitting procedure.

Alignment Studs

Alignment studs are used to guide the oil pump into place during

final installation. Use at least two studs. Place the gasket on the case first, and then thread the studs into the case by hand. They can be made from a couple of long bolts by cutting the ends off and grinding them to a point.

Slide Hammer

The TH400 transmission has two corresponding pump bolt holes tapped for a slide hammer. It is best to use two slide hammers to pull the pump evenly from the case. One hammer works, but you may have to alternate back and forth between the holes in the pump to effectively pull the pump from the case. You can thread a slide hammer into the pump to remove it from the case. Make sure to thread the slide hammer at least five full turns into the pump to avoid breaking off the tool or pulling out some of the threads.

Special Tools

The TH400 transmission doesn't really require any special tools to effectively rebuild it. However, there are a few procedures that can be difficult to complete without them.

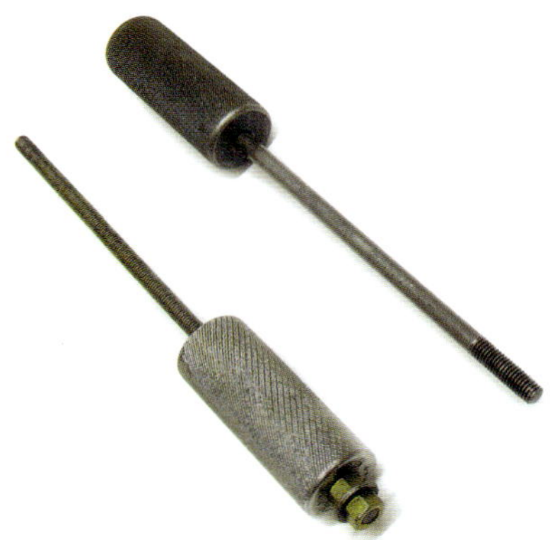

General Motors threaded several bolt holes on the oil pump. By using two slide hammers, the pumps are easily removed from the case without damage.

Valve body accumulators can be difficult to remove and install. Here I have modified a C-clamp into an effective tool for this procedure.

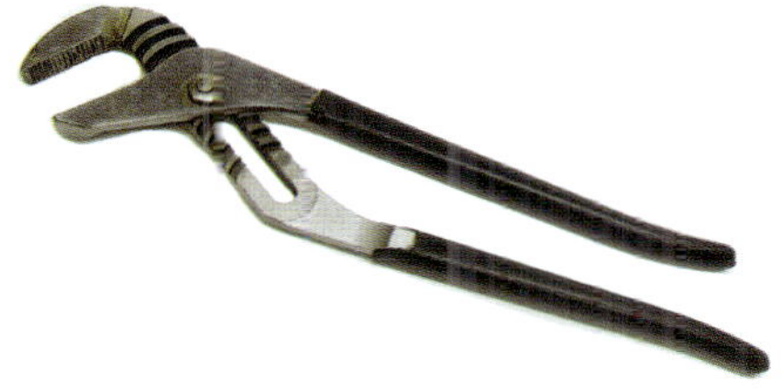

With some patience, a large pair of channel locks also works to remove and replace the TH400 accumulator in the valve body. Make sure to clamp the valve body in a soft-jawed vise, and take care to not damage the sealing ring or bend the pin that guides the accumulator piston.

One is the 2-3 accumulator located in the valve body. You can compress it with a large pair of channel locks in order to remove the E-clip on the pin. Compressing the accumulator for reassembly can be difficult. The accumulator spring is long enough that the sealing ring is outside of the bore before it is compressed. The accumulator piston must be compressed, and the metal ring fitted into the bore, at the same time. I modified a C-clamp to help with this procedure.

A large pair of channel locks works here as well. It may take a third pair of hands to align the accumulator piston with the pin and compress the metal ring at the same time.

Another alternative to channel locks is using a shop press. A shop press can also be used to compress spring cages in order to remove and install retaining rings when rebuilding clutch drums. Although very slight pressure is needed to compress spring cages, a shop press (if you have access to one) is an alternative to making or purchasing spring compressors.

A shop press can be used for clutch pack rebuilding, and to remove the accumulator from the valve body as well. Use extreme care; very slight pressure is required, and spring cages and other parts can be easily damaged if a lot of force is applied to them.

TRANSMISSION FUNDAMENTALS

Automatic transmissions have always fascinated me. I can still remember the very first unit that I took apart more than 30 years ago. It was a General Motors TH400 from an early 1970s Pontiac Bonneville. It

The General Motors TH400 transmission is a strong unit and came in several configurations, including two bellhousing bolt patterns and several different overall lengths. Shown here is a Buick, Oldsmobile, and Pontiac (BOP) unit with a short tailshaft. It also came in a Chevrolet bellhousing bolt pattern version, a long tailshaft version, and even a model with a smaller output shaft. The unit shown is the most common junkyard or swap-meet find, and the TH400 was the factory's choice for full-size cars with large-displacement engines in the late 1960s through the late 1970s.

seemed like the task of removing parts from the case would never end. Even the valve body, which on the TH400 is quite simple compared to most other models, contained dozens of parts. I was afraid that a part or two was going to be left over after the rebuild was complete. I laid each part out very carefully in a long line across the workbench, while paying very careful attention to which direction it had been in. I noted if any thrust washers, snap rings, or other parts were above it, below it, or held the part in the case or inside of another part within the case.

During the rebuild, I gave every part special attention, not only to make sure that I didn't get it upside down or out of place, but also to make sure it was in good serviceable condition and could be reused without causing a malfunction.

My close attention to detail paid off. After many hours of cleaning and then installing new seals, gaskets, sealing rings, and clutch plates, I reinstalled the reassembled unit and it worked flawlessly in all areas.

Although I have built hundreds of these units since my very first experience, I use the same attention to detail and it has always served me well.

Automatic transmission rebuilding is often avoided by highly skilled mechanics and automotive technicians. Due to the complexity of modern, electronically controlled 4-, 5-, and 6-speed units, overhaul and repair has been limited even at dealership levels. Few, if any, automotive repair shops are willing to touch one these days. In almost all cases a complete unit is ordered from a transmission rebuild shop or commercial remanufacturing facility. This is done to save time and money, but does not increase or improve the skills of the automotive mechanic. As with many other items, including smaller items such as starters, alternators, master cylinders, and brake cylinders or calipers, the trend is simply to order a complete unit rather than take the time to completely and correctly rebuild the original factory components.

Many hobbyists, however, are not always concerned with time and efficiency when getting their automotive projects operational. So they have the option to perform many of the repairs themselves.

As far as the automatic transmission is concerned, these efforts save great expense. It is simply one of the areas of this hobby where you can get a better overall end result, learn a great deal about a somewhat complex automotive part, and have the satisfaction of doing it yourself.

Knowing the basic fundamentals of how an automatic transmission works, and the design and function of the internal components, helps in this effort.

Inside of these automatic transmissions, are torque converters, oil pumps, clutch packs, bands and ser-vos, accumulators, planetary gears, sprag and roller clutches, as well as bushings, thrust washers, governors, and modulators. The basic arrangement of the components varies, but the end result of these components working together is the same: They are designed to efficiently and effectively use engine power to propel the vehicle over a wide variety of operating conditions.

Torque Converters

One major component in transmission units is a torque converter, which drives an oil pump located in the front housing bolted to the case. The torque converter is bolted directly to the engine's crankshaft, and transfers all of the energy to the rear wheels. It routes these forces through the transmission's internal components, allows for enough slippage so the vehicle can remain at a stop at idle speed, and is efficient enough to provide very close to com-

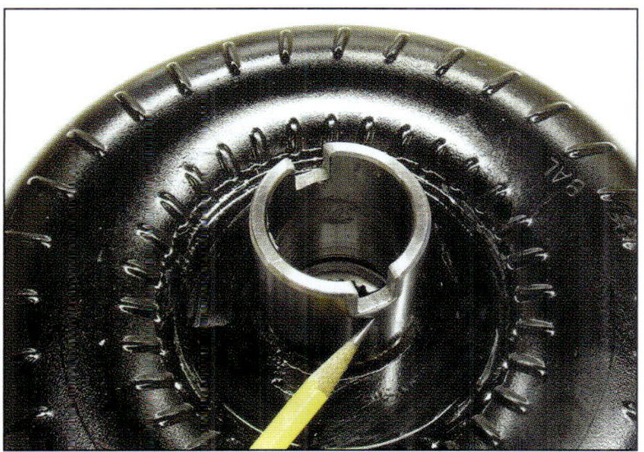

The torque converter is bolted directly to the engine's crankshaft and spins at engine speed. The hub on the converter has two notches that engage with the transmission's oil pump to provide constant oil flow and pressure. The oil pump is the heart of the transmission, supplying oil for cooling, lubrication, and to operate the various internal components for transmission function.

plete energy transfer at various vehicle speeds. Torque converters equipped with an internal clutch can actually be employed once the vehicle is in motion, to provide a complete or solid lock between the engine's crankshaft and the transmission's input shaft, just like the clutch does in a manual transmission.

The torque converter serves double duty; it turns the transmission's input shaft and drives the oil pump at the same time. This provides oil flow and pressure at all times while the engine is running. Some oil is diverted to the transmission's oil cooler, some is used to keep the torque converter full, and the rest is used to supply pressurized

The torque converter is used to provide a coupling between the engine and the transmission. It is also used to drive the transmission oil pump. Since the torque converter spins at engine crankshaft speed, the outer housing is always rotating when the engine is in operation. The design of the internal components allows the vehicle to come to a full stop, and the engine to continue to operate without stalling. When the operator presses down on the accelerator pedal, the vehicle moves out and continues to accelerate while effectively using engine power converted to motion by the fluid inside the torque converter.

---

oil to various components within the transmission for lubrication and basic transmission function.

The torque converter is used to provide the driving force for the transmission. It is attached directly to the engine via a flexplate. The hub of the torque converter is notched to drive the transmission's oil pump. This provides constant oil flow and pressure any time the engine is running.

The internals of the torque converter allow the vehicle to stop and the engine to continue to operate without having to disengage a clutch. This is due to the internal design, which is basically arranged like two fans facing each other. The blades on the fans do not touch, and have a very minimal distance between them. One fan is mounted directly to the converter housing and moves fluid toward the other fan, which is attached to the transmission's input shaft. At low speeds, or at nearly idle engine speed, there is enough distance between the two fans to allow one to turn without moving the other. But as engine RPM increases, so does fluid flow. The opposing fan, which is connected to the transmission's input shaft, reacts from the force and flow of the incoming fluid and begins to turn the input shaft. Another component, called a stator, is used to increase the efficiency of the fluid moving inside the torque converter.

The drive fan, which is also called an oil pump or impeller, contains a series of vanes. Because it is mounted to the converter shell, it is always moving oil when the engine is in operation. The amount of oil moved is directly proportional to engine speed.

The turbine is mounted directly to the transmission's input shaft. It uses a series of vanes very similar to the impeller. Because the oil from the impeller is moving toward the turbine, it reacts to the force and flow of the oil. The more oil that is moved toward the turbine, the harder it tries to turn the transmission input shaft. It basically converts the flowing oil from the impeller into mechanical energy.

The stator is mounted between the pump and turbine and is used to further increase the efficiency of the oil that is moving within the torque

The stator assembly is splined directly to the stationary input splines on the front of the transmission. It is located between the impeller and the turbine, and is a key player in torque converter efficiency. The stator vanes are used to redirect some of the oil flowing from the pump back to the turbine, instead of allowing it to flow away from the turbine and not use the energy that it contains to drive the input shaft.

Either a one-way sprag or a roller clutch is used in the stator assembly. The clutch is on, or applied, at low engine and vehicle speeds, when the most slippage between the turbine and impeller occurs. With the stator being held stationary to the stator shaft on the transmission, it holds the entire assembly from turning, redirecting oil back to the turbine. As the engine and transmission input shaft speeds achieve approximately the same RPM, the stator clutch releases.

The torque converter pump or impeller is built into the rear portion of the outer shell. It uses a series of vanes to move transmission fluid, much like a fan moving air away from it in a room. Since the fluid being moved is oil, it is used to transfer mechanical energy created by the engine to the transmission's input shaft.

The turbine inside the torque converter is splined directly to the transmission's input shaft. It is mounted directly in front of the impeller and is constantly receiving oil when the engine is in operation. The fins or vanes on the turbine react to the oil being moved against them, turning the oil flow into mechanical energy.

converter. A one-way roller clutch or sprag is located inside the stator assembly. It is on and holding at low vehicle speeds, and re-routes some oil back to the turbine. This improves its efficiency and torque multiplication. Eventually the impeller and turbine achieve approximately the same speed, and then the one-way clutch releases.

Transmission Oil Pump

The transmission oil pump is driven directly from the rear snout of the torque converter. Since the torque converter is bolted directly to the engine's crankshaft, the oil pump operates at engine speed. The oil pump uses a pressure relief valve to keep oil pressure within a specific range for best transmission operation. Oil from the pump is routed through the vehicle's transmission oil cooler because heat is created in the transmission and inside the torque converter. Oil is also routed to various transmission internal com-

ponents through the valve body at various gear-selection sequences to help keep the engine in its most efficient speed/load range over a broad variety of driving conditions.

General Motors used two different types of oil pumps. Early models, such as the TH400, used a gear-type pump. The internal gear is made so that it engages with two cutout slots on the end of the torque converter hub. The outer gear is offset inside the pump to provide fluid flow and pressure as the pump turns. The pump tolerances are very close within the pump halves to maximize efficiency.

Clutch Packs

Another feature common to all of the General Motors automatic transmissions is the clutch pack. The number of clutch packs used, and their arrangement varies with the design of the particular unit, but they are all used in (and function in) a similar manner.

Clutch packs can be in the transmission case, or in a separate drum within the case. A clutch uses multiple plates, both steel and friction lined. They are sandwiched together in a confined space. The steel plates have outer splines attaching them to one part, such as the transmission case. The friction plates have inner splines attaching them to a moving part within the case. An apply piston is used to apply force to the clutch pack to stop the rotation of one or both parts, or to stop the motion of a part moving within the case when the steel plates are splined directly into the case itself.

The steel and friction plates are immersed in a constant bath of transmission fluid. This helps cool

TH400 transmissions use a gear-type oil pump. The inner gear is driven directly by the hub of the engine's torque converter. Note that the outer gear is offset in the housing. This creates a chamber that moves oil and forces it out of the pump.

Within the TH400 direct clutch pack is an apply piston with seals, spring cage, snap ring, steels, frictions, backing plate, and retaining ring. Pressurized oil is routed behind the apply piston, forcing it to move toward the steel and friction plates. Note that the steel plates are notched on the outer edges, while the frictions have internal teeth. When the clutch pack is fully compressed, it effectively locks the drum to the component that has splines engaged with the frictions. In this case the component is the forward clutch, so both components turn at the same speed.

A pressure regulator (PR) valve is used to control transmission oil pump output pressure. The TH400 pressure regulator valve is located on the back side of the oil pump. Some PR valves are located in the transmission's valve body instead, but function in the same manner. The oil leaving the oil pump must overcome spring pressure to maintain flow to the rest of the unit.

Inside a clutch pack are frictions and steel plates. The basic arrangement is to start with an apply plate, which in many cases is a steel plate, then alternate friction–steel–friction–steel, and end with a backing plate. This basic arrangement of parts is used for all automatic transmission clutch packs. Keep in mind that the apply plate may be a solid steel plate pushing directly on a friction plate, or it may push on a steel plate first. This is an area of transmission building that gets considerable attention because there may have been several different combinations of parts used depending on the transmission model and application.

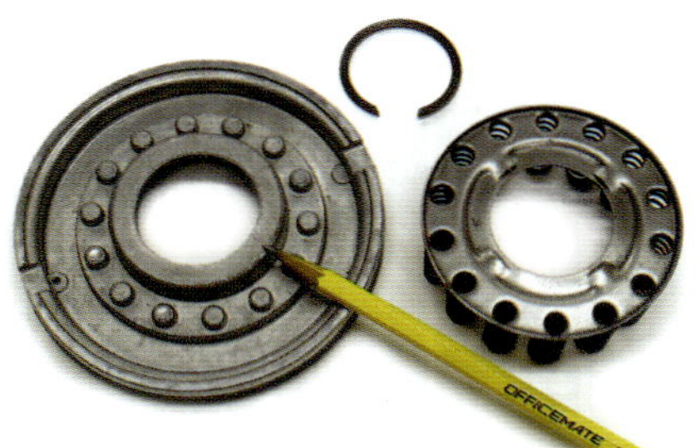

An apply piston is used within a clutch drum to compress the confined steel and friction plates. Pressurized oil is routed behind the piston, causing it to move and apply force to the clutch pack. The clutch pack must overcome spring pressure. A spring cage, held in place with a snap ring, is used to move the piston away from the clutch pack when the fluid pressure is removed from behind the piston.

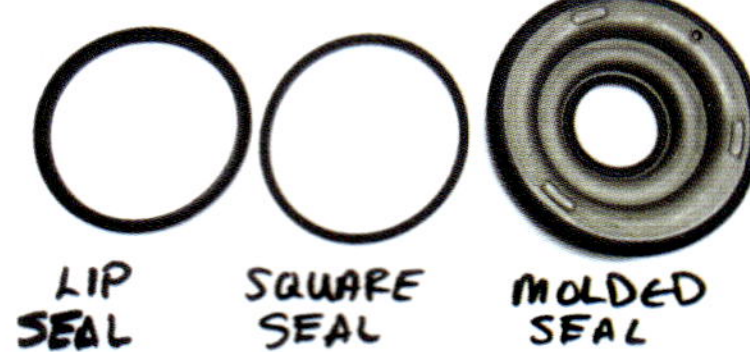

Several different types of seals are used on apply pistons. The most effective and easy to install are the molded pistons (right). Many late-model transmissions adapted molded pistons where the seals are a permanent part of the piston. This requires replacing the entire part during rebuilding, in order to renew the seal material. Lip-type seals (left) are also very common and very effective at sealing off incoming transmission fluid. The pressurized fluid forces the lip seals tightly against the housing as the piston is applied. Square-cut seals (middle) are also used and relatively easy to install. They are typically used where access to the apply piston is limited. This makes the piston easy to install into the case without special tools or procedures.

A spring cage is used to keep the apply piston away from the clutch pack. When pressurized transmission fluid is routed behind the apply piston, it moves toward the spring cage (overcoming the spring pressure) and forces the clutches and steels tightly together. The moment the pressure is released, the spring cage pushes the apply piston away from the compressed clutch pack.

them, minimize wear, and ensure smooth operation.

As far as the rebuilder is concerned, clutch packs share common parts and orientation of the parts. The apply piston uses seals to prevent leakage as pressure is applied to it. Some pistons use lip-type seals, some use square seals, and others have molded seals. The piston is held away from the clutch pack by springs, often contained in a spring cage, which is retained by a snap ring. As the piston moves toward the clutch pack, it contacts an apply plate. The apply plate can be one of the steel plates in the clutch pack, or

A backing plate is used in a clutch drum, behind the last friction plate. This provides a positive stop for the compressed clutch pack when the apply piston forces the steels and frictions tightly together. Backing plates are typically very thick and sturdy parts, and may be selective (or different) thicknesses. They may also interchange within different drums in the same transmission, so make sure you return them to their original location during the rebuilding process. A large snap ring is used to hold the backing plate in the drum and to keep it in place when the clutch pack is compressed.

Some clutch packs contain a waved steel plate. The waved plate flattens out as the clutch pack compresses. This delays or softens engagement some, and is often used to minimize customer complaints. A common location for a waved steel plate is in the forward drum, for example. The waved plate may be used between the apply piston and a flat steel plate, or may be used directly against a friction plate. Some builders replace the waved plates with flat steel plates during rebuilding, to firm up clutch pack engagement and shorten shift timing.

Steel and friction plates may come in several thicknesses. Shown here are direct and forward steel and friction plates used in the TH400 transmission. The thicker steel plates were used in the direct drum, while the thinner steel plates were used in the forward drum. This may not always be the case, but is the general rule for this particular transmission. It is quite common to find thick and thin steel plates in the same drum, or the thin plates may be used to add frictions, thus creating more space.

rations were used, you must pay close attention to the arrangement of the parts when taking the clutch pack apart for inspection/replacement.

Steel and friction plates are available in several thicknesses. This allows you to custom-stack the clutch pack in the drum and establish the correct endplay. You can also use thinner materials to add more friction plates to the clutch pack, to increase the surface area and durability of the unit.

Bands and Servos

Bands are used in transmissions much like a clutch pack, except they have a single layer of friction material or lining on the band. They are arranged so they grab and hold a particular component and keep it from turning. One end of the band must be held stationary, usually by a "notch" in the case. The other end of the pan engages with an apply piston, usually by means of a pin. The apply piston is called a servo, and works in much the same manner as an apply piston in a brake drum. The servo has a return spring and retainer to quickly disengage the band when the hydraulic pressure is removed.

A servo is nothing more than a small piston used to apply a band within a transmission. Bands are used in nearly every General Motors transmission, because they take up much less space than a clutch pack. As with a clutch pack apply piston, pressurized hydraulic fluid is used to apply the servo. The servo has a pin or strut connected to one end of the band. The movement of the servo causes the band to grab whatever component is under it. It stops that component from spinning and holds it tightly until the fluid pressure is removed from the servo. A return spring is under the servo to move it away from and release the band.

One advantage to using a band in a transmission is that it takes up much less space than a drum and multiple-disk clutch pack. The downside of this is that it has much less combined surface area than a multiple-disk clutch pack, and is asked to stop a transmission component that may be spinning at a relatively high RPM. The servo timing and apply force must be well calibrated for the band to provide a smooth shift, yet not soft enough that the lining on the band becomes too hot and burns/cracks.

a separate and sturdier (thicker) metal plate. The clutch pack, from this perspective, is a series of friction plates alternating with steel plates. The last friction plate has a thick metal backing plate behind it, held in place by a snap ring.

Some clutch packs use a waved or beveled steel plate to help cushion the apply force to the clutch pack. In some cases the waved plate may be in direct contact with a friction plate; in other cases a steel plate and then a friction plate. Since different configu-

Accumulators

Automatic transmissions use hydraulic fluid under high pressure to apply bands and clutch packs. The size of the servo (the total surface area that is pushed on by the incoming transmission fluid under pressure) determines the apply force applied to the band or apply piston. This force can be considerable, and can cause the servo or apply piston to react and activate the band or clutch pack quite quickly and with great force. To help soften the apply force, an accumulator may be used. An accumulator is nothing more than another piston with a spring behind it.

Just as the servo or apply piston moves away from the flow and pressure of the incoming transmission fluid, so does the accumulator piston. The diameter of the piston and the tension of the spring behind it are precisely tuned to allow the component to engage without creating an overly hard or firm shift. They also minimize any potential for "fluid shock," which is similar to what happens on occasion in the pipes of household plumbing when a valve is opened.

Basically, the accumulator piston reacts in conjunction with the apply piston and continues to compress the spring behind it as the apply piston applies the clutch pack, or as the servo applies the band to the drum under it.

Some accumulators also have a portion of the applying transmission fluid routed in behind them to assist the spring in slowing the movement of the piston. This further reduces the possibilities of fluid shock and is carefully calibrated to provide a smooth shift with the transmission.

Many aftermarket shift kits recommend, and even supply, components to eliminate accumulators and their function. This configuration applies all of the fluid directly to the clutch packs' apply piston or band servo and, in most cases, considerably firms up that particular shift. It also speeds up the shift, sometimes so much that the shift applies with such force as to actually "chirp" or spin the vehicle's rear tires! This can be incredibly hard on transmission internals. Better-quality shift kits typically provide stronger springs to use behind the accumulator piston. And in some cases a fitting is provided to drive into the transmission case, in order to limit or slow the fluid flow that may be routed behind the accumulator.

Planetary Gears

In an automatic transmission, planetary gearsets are used to provide different gear ratios and constant torque flow from the engine to the drive wheels. They are called "planetary" gears or gearsets simply because of the basic design of the components. A "sun" gear is the

Accumulators are used to help control shift timing and shift "feel." They are placed in line to the clutch drum or to the servo that applies a band. Pressurized fluid headed to the clutch pack apply piston or servo pushes on the accumulator piston. At the same time, the fluid also moves the apply piston to compress the clutch pack or the servo to apply the band. Because the fluid must also compress the spring under the accumulator, it delays or lengthens the shift timing slightly. In some transmissions the fluid may also be routed behind the accumulator through a very small passage, or a passage with a restricted orifice in it. This assists the spring in holding back the accumulator.

Shift Improver Kits

It has long been the practice of many transmission specialists to disable accumulators and enlarge separator plate holes in order to provide the customer with much firmer shifts. Instead, I recommended purchasing a shift kit specifically designed for this purpose. The companies that make shift kits carefully calibrate their kits to provide an ideal combination of fluid flow, pressure, and accumulator function. This optimizes shift performance without being unnecessarily hard on the transmission's internals.

innermost part of the planetary gear. The planetary gears are held at fixed positions, evenly spaced around the sun gear in a planet carrier, and surrounded by a ring gear.

The planetary gearset has significant strength advantages when compared to two meshed gears in a typical manual transmission. This configuration provides a more even application force and load distribution, and increased contact area. Most units use either 3- or 4-pinion planetary gearsets with helical-cut gear teeth. The helical cuts on the teeth further increase tooth contact surface area when compared to straight-cut teeth. The negative to

Planetary gearsets are used in automatic transmissions to provide various gear ratios. The center (or sun) gear is meshed with planetary gears held in a carrier. These are also in mesh with an outer gear or carrier. Any of the three parts can be held or driven to provide different gear ratios and/or functions. If the outer ring gear is held stationary and the sun gear is driven, the planetary gears rotate around the sun gear and rotate the planet carrier in the same direction as the sun gear, but at a slower RPM. This provides gear reduction. Holding the sun gear and rotating the ring gear results in the planetary gears rotating around the sun gear, but with less gear reduction. Holding the sun gear and rotating the planet carrier results in the ring gear turning with gear reduction in the opposite direction (reverse). Direct drive (a 1:1 input/output ratio) is achieved by locking any two elements of the planetary gear set at the same time.

A sun gear is the innermost gear in a planetary gear set. The planetary gear pinions revolve around the sun gear and within a ring gear. This provides three separate elements, which can be held or driven to provide gear reduction, reverse, or a 1:1 (direct) drive.

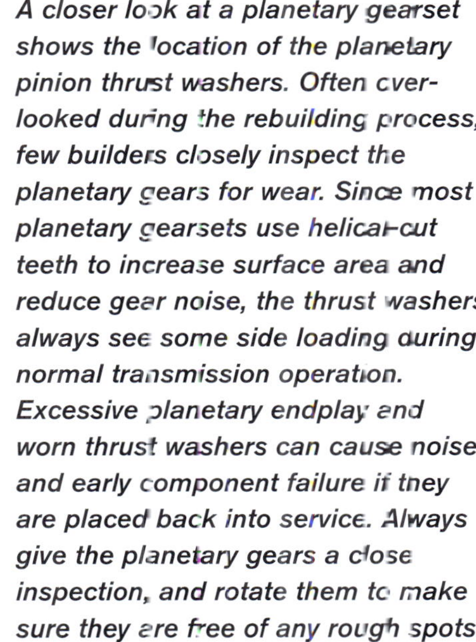

A closer look at a planetary gearset shows the location of the planetary pinion thrust washers. Often overlooked during the rebuilding process, few builders closely inspect the planetary gears for wear. Since most planetary gearsets use helical-cut teeth to increase surface area and reduce gear noise, the thrust washers always see some side loading during normal transmission operation. Excessive planetary endplay and worn thrust washers can cause noise and early component failure if they are placed back into service. Always give the planetary gears a close inspection, and rotate them to make sure they are free of any rough spots.

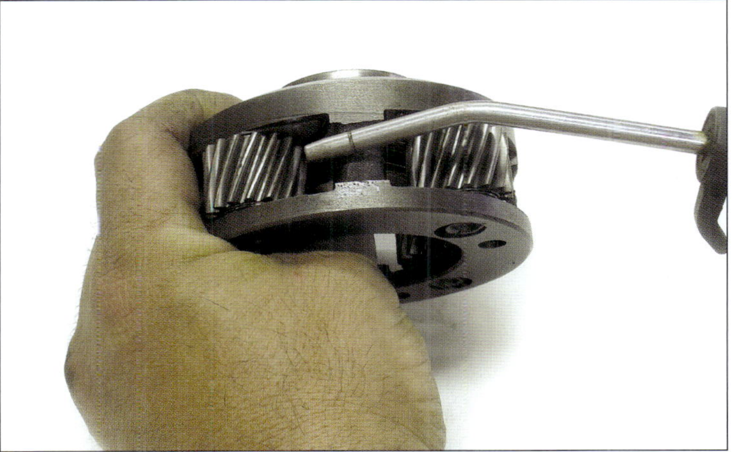

Avoid the temptation to spin the planetary pinions with a high-pressure air gun. This can quickly damage the tiny needle bearings under the pinions and the thrust washers.

Planetary gears should be completely submerged in transmission fluid for several minutes prior to installing them to make sure all of the components are well lubricated when the transmission is first placed into service.

---

helical-cut teeth, when compared to straight-cut teeth, is that more side loading occurs against the thrust washers.

Planetary gears and their thrust washers should be very carefully inspected during any transmission rebuild. In almost all cases, the teeth surface area is fine, and they spin freely on the sun gear and within the ring gear without any binding or rough spots. Even so, they may have excessive endplay and heavily worn thrust washers. With close inspection you can usually see when the brass thrust washers are heavily worn. In some cases they are completely worn out, which allows the planetary pinions to ride directly against the steel backing washers. A great deal of common sense must be applied here; if the parts seem excessively loose, compare the planetary gearset to a new one. You can also consult factory tech publications for the maximum allowable sideplay, and then measure and replace them as needed.

Avoid the temptation to spin the planetary gears with compressed air. This can be a dangerous practice. It not only damages the gears and the needle bearings under them, but they can fly apart with great force much like a small grenade! After the planetary gears are thoroughly inspected and cleaned, submerge them in transmission fluid for at least 10 minutes prior to installation.

Sprags and Roller Clutches

Sprags and roller clutches are used in automatic transmissions as an alternative to bands or multiple-disk-type clutches. They take up very little space, and are very strong. They are not hydraulically actuated; they are a simple mechanical device, and very similar to the ratchets used to drive sockets. They allow rotation in one direction, and lock solidly in the other direction. They also apply pressure very quickly compared to fluid-operated clutch packs or bands.

There are basically two types: sprag, and roller. The roller-type clutch uses a series of rollers and springs in a notched housing. The

A sprag-type clutch operates in a manner very similar to the roller-type clutch. It uses dog-bone-shaped internal components instead of rollers. Both the inner and outer races are smooth, and there are no notched areas as found in a roller-style clutch. The internal elements are long enough and shaped so that one race can turn independently from another in one direction. When turned in the opposite direction, the unit locks solidly. Multiple-element sprag assemblies, such as the 34-element intermediate TH400 sprag shown, are considered to be a high-performance upgrade over the roller-style clutch typically used in that location.

Most TH400 transmissions used a notched direct drum that holds a roller-clutch assembly. Very early units had a smooth drum and outer race and used a sprag-type clutch. A 34-element sprag is available from the aftermarket, or one can be used from a 4L80E transmission. This is considered a high-performance upgrade for any TH400 transmission that originally came with a roller-type intermediate clutch.

The TH400 smooth and notched intermediate drums are shown here. The sprag-type clutch is used in the drum with a smooth inner race. These drums are becoming difficult to obtain in good used condition. You can substitute a 4L80E drum; they are available new through most transmission parts suppliers.

other race is smooth, and rotates easily in one direction. If the direction is reversed, the spring-loaded rollers move back toward the notched areas of the inner race. Since they are too large in diameter to fit through the openings, they quickly and effectively lock the unit. Roller clutches are smooth and fast to apply, but lack the surface area and overall strength of sprag-type clutches.

A sprag-type clutch simply uses a series of dog-bone-shaped, roller-bearing-type components held in a stationary position between a smooth inner and outer race. All of the sprags are lying in one direction between the races, and allow free rotation of the outer race in one direction only. If the direction is reversed, the sprag elements are too tall to allow the outer race to turn as they try to stand straight up within their confined space. Sprag-type clutches may contain two to three times as many elements as roller-type clutches. This not only makes them very strong, but they also distribute their load over a greater surface area. For this reason, high-capacity sprag-type clutches are preferred for high-performance use. A good example of this is the 34-element intermediate sprag assembly in the TH400 and 4L80E transmissions.

Many TH400s use a roller-type clutch at this location instead of the sprag-type clutch. An upgrade for the TH400 is to install a 4L80E drum and sprag assembly, or obtain a smooth drum from an early TH400 transmission. Most units made prior to 1973 have the smooth drum and have used a sprag-type clutch, although it was not a high-capacity unit. The 4L80E sprag uses 34 elements, and the 34 element sprag is available separately for upgrading any TH400 transmission that uses a smooth drum.

High-capacity sprag-type clutches are available for building a high-performance transmission. The TH400 transmissions that used a smooth direct drum can be upgraded by installing the 34-element sprag assembly. You can also use a 4L80E drum if a smooth TH400 drum is not available. I also recommend installing the 4L80E spiral lock ring to hold the sprag assembly in place.

Valve Body

The valve body is basically the brain of the transmission. It contains an assortment of valves, which include the manual valve, shift valves, and the throttle valve. It may also contain the pressure regulator valve, TCC lock-up valve, and one or more accumulators. The valve body

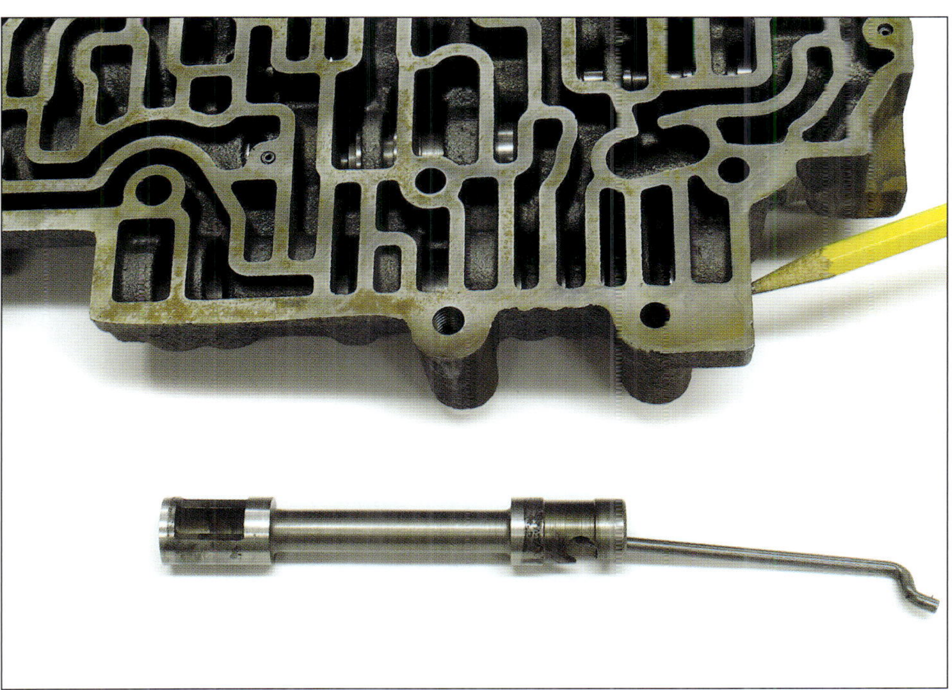

The manual valve is connected directly to the vehicle's shift selector. It moves in the valve body to provide the correct fluid path for forward and reverse operation, as well as manual operation of the gear range selected by the operator.

Roller clutches are used in many General Motors transmissions. The rollers are confined within notched areas and held in position with small springs. This allows the assembly to rotate freely in one direction. When the direction is reversed, the rollers move within the notched areas and effectively lock the inner and outer races together. Roller clutches are very strong, but lack the total surface area of engagement found with sprag-type clutches.

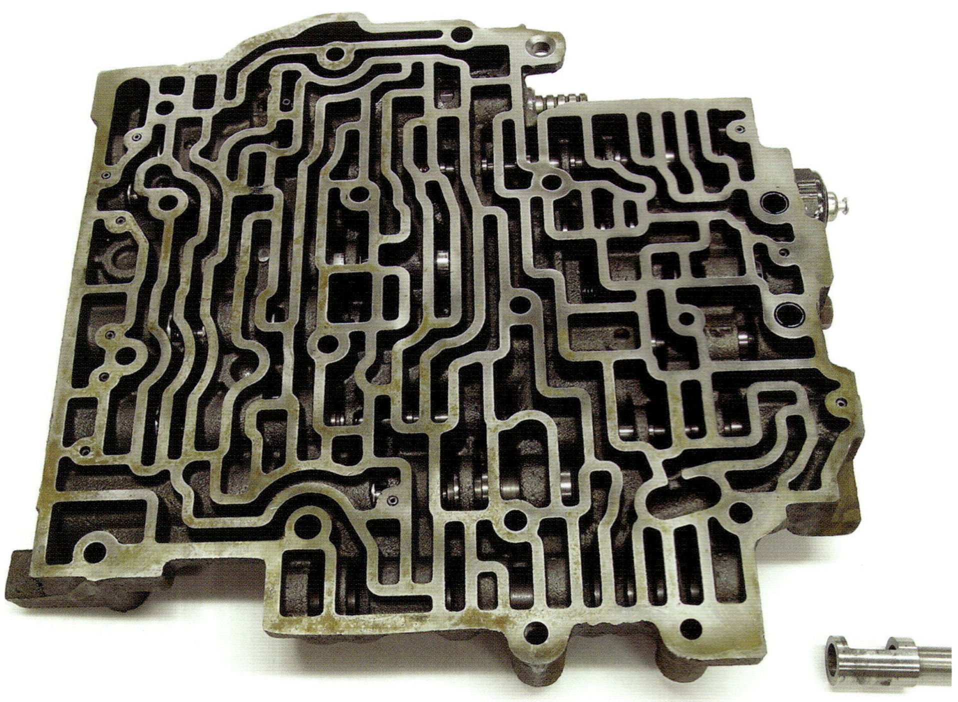

The valve body often appears quite complicated at a glance, with dozens of internal passages, valves, check balls, and springs. The valve body routes fluid flow to the correct transmission components to provide constant vehicle motion through gear changes, while effectively transferring engine power to the rear wheels. Before engine management computers took over transmission shifting functions, the valve body received signals from several mechanical components to help it select the appropriate gear for the driving situation.

Many early transmissions, such as the TH400, used a vacuum modulator mounted to the transmission case. The modulator references engine manifold vacuum and provides the appropriate signal to the transmission as it relates to engine throttle position and engine load. Low-vacuum situations, which require heavy-throttle or full-throttle application (such as when passing another vehicle), provide a low-vacuum signal to the transmission via the vacuum modulator. This results in delayed upshifts, allowing the engine to rev to a more appropriate RPM for faster vehicle acceleration. Very light throttle operation, as another example, provides a high-vacuum reading to the vacuum modulator, resulting in quicker transmission upshifts at lower engine RPM. Other devices, such as a transmission downshift solenoid (kick-down valve), were also used in some models to force gear changes to lower gears for heavy- or full-throttle situations.

The transmission's governor uses springs and weights and is driven by the output shaft. This references vehicle speed, and the governor is carefully calibrated to provide the valve body with the correct reference signal for effective gear changes.

is tied in hydraulically to the governor on transmissions that are not computer controlled. It may also reference engine vacuum via a vacuum modulator located in the case that is hooked to an engine manifold vacuum source.

The governor is located in the rear portion of the transmission, and driven by the transmission's output shaft. Flyweights and springs are used and calibrated to provide the appropriate signal to the valve body for controlling gear changes.

Some transmissions also use a vacuum modulator, which is mounted to the transmission case. It is hooked to an engine manifold vacuum source to provide a reference signal to the transmission. It is also calibrated to assist with gear changes.

By sensing vehicle speed, engine vacuum, and throttle position, the valve body is calibrated to provide full automatic shift function through all of the forward gears. The manual valve provides the vehicle operator with manual gear selection, which may override automatic shifting. When the vehicle gear position selected is in a gear lower than the transmission's highest gear, such as when it's placed in second gear (or "2," for example), the transmission shifts normally from first to second gear, but stays in second unless the shift selector is moved.

Some transmissions, such as the TH400, allow automatic upshifts to second gear, even if the vehicle gear shift selector is placed in first (or "L"). This occurs at very high RPM. Most transmission shift kits allow the installer to eliminate this function and provide the operator with full manual control of the unit using the shift selector, but, otherwise, it retains its normal automatic function.

Some factory transmissions (such as the TH400) were designed to upshift even if the shift selector is left in the "L" (or first-gear) position. This only occurs at high engine RPM and is basically a safety device to protect the engine from overreving. This is great for a low-performance daily-driven car, but not the best for racing or a high-performance application. Some shift kits contain a modified 1-2 shift valve, which eliminates the automatic upshift function and provides the operator with full manual control of first and second gear. Keep in mind that, if such a valve is installed, the transmission now shifts into low gear at any vehicle speed, when the manual low gear position is selected.

TRANSMISSION REMOVAL

Removing and installing transmissions can be difficult undertakings. The vehicle needs to be lifted high enough to get the transmission in and out from underneath. This isn't too difficult with pickup trucks because they are pretty high off the ground to start with. For most cars, the vehicle must be lifted and well supported.

The engine needs to be supported as well; it uses the transmission and rear transmission mount for support. The engine's exhaust system supports it to some degree, but should not be trusted by itself to keep the engine from tilting backward when the transmission is unbolted from it.

Some engines have rear-mounted distributors, and the distributor cap may be close to the engine compartment firewall. If the engine is allowed to tilt backward, it could break the distributor cap, and even bend or damage the distributor housing.

The best-case scenario for transmission removal and installation is to have a vehicle lift and a transmission jack. Most hobbyists do not have these items, but in recent years two- and four-post lifts have become very affordable, and lots of folks are installing them in their garages and workshops. They allow excellent car storage when not in use. Most require around 12 feet or so of clearance to the rafters to lift most cars high enough to work on them.

In my own shop, I had about 12½ feet of clearance to the bottom of the trusses. This was fine for cars, but did not allow the cab of a truck to be high enough to get underneath to work on it. With a few minutes of work I was able to modify one truss to provide enough room for the cab. Doubling up on each side and spreading the load to the two adjacent

Shown here is a four-post lift. Having the vehicle on a lift makes transmission removal and installation much easier. The lift provides a stable support and you can stand under it. I still remember the days of doing these jobs in my driveway on my back! The lift was one of the best purchases I ever made, and I still can't believe that I worked so many years on these things without one!

trusses makes sure the roof can still provide adequate support.

I also modified the transmission jack to provide additional lift when needed. A transmission jack is just about mandatory when using a lift and standing under the vehicle. You can work off the ground with floor jacks, but even then a transmission jack is much better. Floor-type transmission jacks provide additional lift capabilities in comparison

Here the vehicle has been raised slightly and the engine oil pan has been lowered onto a couple of blocks of wood. This serves to support the engine when the transmission is removed. In most cases the vehicle's exhaust system effectively supports the engine, but it can still tilt back some. Some applications also require that the exhaust system behind the manifolds or headers be removed to have sufficient clearance to remove and replace the transmission. This mandates some form of support for the engine.

Here's a common problem with some vehicles with a rear located distributor. If the engine isn't effectively supported, it can lean back and break the distributor cap and, in extreme cases, bend or break the distributor housing! Be mindful of how close the distributor is to the firewall.

A transmission jack is essential when using a lift for transmission removal and replacement. It holds it in place for hours and allows the transmission to be lowered slowly to get past obstacles that may need to be moved on its way in and out of the vehicle.

It takes a lot of clearance to lift up large trucks and SUVs. I had to modify one truss in the shop for cab clearance when lifting larger vehicles, even though there was 12½ feet of clearance under the trusses.

Large trucks and four-wheel-drive vehicles with additional ground clearance may require more height capabilities. I have added an extension to the transmission jack accordingly.

to a standard floor jack. They have more room for travel and a wider base. You can strap the transmission to them, and also tilt it into the opti-mal position during removal and installation.

After you have your vehicle supported and the safety jacks installed, you can begin removing the transmission. It is always best to drain the transmission fluid first, to avoid any spills out of the tail housing area or through the filler tube hole if it is removed. Remove the battery ground strap to prevent accidental

1 Remove Pan Bolts

If the transmission is not equipped with a drain plug, it can be drained by removing the transmission pan. It is best to leave two bolts in place on one end of the pan or the other, and allow it to tilt downward to drain most of the fluid from it.

2 Remove Ground Strap

The ground strap should be removed from the battery to make sure the engine can't be started while the transmission is out of the vehicle.

3 Remove Starter

The starter may need to be removed to access the torque converter bolts. Doing so allows plenty of room from the front side to effectively torque them down once the transmission and torque converter are back in place. In most cases, the starter wiring can be left in place, and the starter can hang from a coat hanger or heavy wire during this procedure.

engine starting during this procedure. The engine starter may also need to be removed from the engine.

If you can obtain a spare yoke, you can insert it into the tail housing when you remove the driveshaft. When removing the driveshaft, make sure to take extra care not to lose the caps on the U-joint. The tiny needle bearings are easily dislodged from the caps if they hit the floor. A couple of wraps of electrical or duct tape keeps things together until the shaft is reinstalled and ready to be reconnected to the differential yoke.

Professional Mechanic Tip

4 Use Spare Yoke

If the transmission fluid isn't drained, a spare yoke can be used to keep fluid from leaking out the rear of the unit.

5 Tape Over Rear U-joint

As soon as the driveshaft is removed, wrap the rear U-Joint with tape to keep things intact. There is nothing worse than having one of the ends fall on the floor and all the needle bearings come out of it!

Unbolt the torque converter from the flywheel and keep it in the transmission while lowering it. Remove the torque converter inspection cover to access these bolts. In most cases, the engine can be turned over by hand to gain access to the torque converter bolts. Use a big screwdriver or pry bar and turn the ring gear. If not, the engine crankshaft can be turned with a deep-well socket and ratchet. The engine has to be turned one full revolution to access all of the torque converter bolts. TH400 transmissions use torque converters with three bolts attaching them to the flywheel.

6 Remove Dust Cover

Remove the dust cover to gain access to the torque converter bolts.

7 Turn Engine

A big flat-tip screwdriver or pry bar can be used to turn the engine to gain access to the converter bolts because some applications have limited access to the front crankshaft bolt from the underside of the vehicle.

8 Remove Torque Converter Bolts

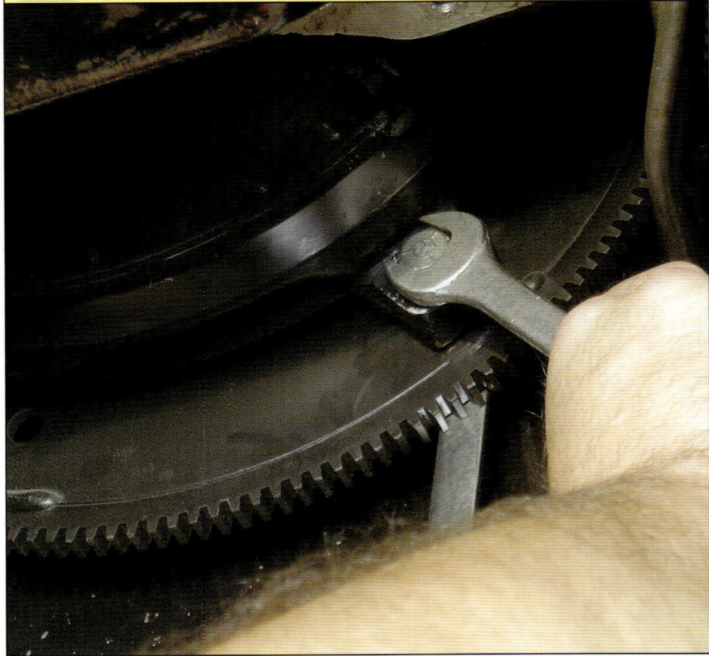

Remove the torque converter bolts. Many General Motors applications use bolts that thread into the converter. Shown here is an aftermarket converter using larger-grade-8 $\frac{7}{16}$-inch bolts and nuts.

Remove any shift linkage, downshift linkage/throttle valve linkage, and transmission cooling lines. Locate and remove the speedometer cable as well. Remove any electrical connectors that are attached to any plugs on the transmission case. It is best to pull these items out of the way and secure them with wire, wire ties, or tape to keep them from gettingdamaged when the transmission is lowered from the vehicle. Loosen and remove the transmission cooling lines. It may be necessary with some applications to lower the transmission slightly to gain access to the cooling lines.

9 Remove Shift Linkage

Remove the shift linkage where it attaches to the transmission. Some applications also have a bracket under two of the pan bolts.

10 Remove Speedometer Cable

Remove the speedometer cable with a pair of channel locks or large pliers.

11 Remove Electrical Connections

TH400s used an electrical solenoid for downshifts. Remove any electrical connections to the transmission, and safely secure the wiring out of the way.

12 Remove Cooling Lines

Cooling lines can be difficult to access, and several different types of fittings are used. You must hold the transmission fitting when removing the lines. Most General Motors TH400s used 5/16-inch inverted flare cooling lines. With some applications, it may be necessary to lower the transmission slightly to gain access to the cooling lines. Make sure not to bend or kink them during this procedure.

If working on a four-wheel-drive vehicle, you have two driveshafts to remove, and a transfer case attached to the back of the transmission. It is best for some applications to remove both the transmission and transfer case as a single unit; accessing the bolts between the two components can be very difficult. If you have the means to effectively support the weight, and balance the two parts, I recommended that they be kept together for removal and installation.

13 Remove Driveshaft

Four-wheel-drive applications require removal of the front and rear driveshafts. The transmission is bolted directly to the transfer case, which supports the transmission. They can be removed as a single unit, or separated and removed individually, depending on the resources available. If the transfer case is removed first, the engine needs sufficient support to hold the weight of the transmission.

There are also additional components, such as shift linkage and wiring, speedometer cable, etc., involved with the transfer case. Make sure everything is disconnected before lowering it from the vehicle.

Raise the transmission slightly and remove the rear crossmember. On some units it also helps to remove the rear transmission mount to provide additional clearance to get the crossmember out of the frame rails. The transmission jack must be in place at this time to support the unit.

14 Remove Crossmember and Rear Mount Bolts

Two bolts attach the transmission's rear mount to the crossmember. After you remove the crossmember and rear mount bolts, you can raise the transmission slightly so the crossmember can slide out from under the transmission.

The transmission bellhousing bolts are the last items to remove before lowering the unit. The filler tube can stay with the transmission with some applications. With others it has to be removed first, depending on the clearance between the back of the engine and the firewall area.

15 Remove Bellhousing Bolts

Six bolts attach the transmission to the engine. It is best to remove the upper bolts first, leaving the two bolts just above the engine dowel pins for last.

16 Use Universal-Joint Socket

Some upper bellhousing bolts can be very difficult to access from the underside of the vehicle. Gaining access to them may require a long extension, or several combinations of extensions and universal joints.

17 Remove Filler Tube (if applicable)

The filler tube is held stationary into the transmission by one of the upper bolts on the right side of the transmission. In some applications, you need to remove the filter tube before removing the transmission.

18 Use Dowel Pins

Two large dowel pins, one on each side of the engine, keep the transmission in perfect alignment.

The transmission bellhousing is supported by two large dowel pins. They help to align and support the engine and transmission and are pressed into the back of the engine block. The transmission must go back slightly before lowering it. It only needs to go back far enough to allow the hub on the torque converter to clear the engine crankshaft flange.

Some vehicles also need one or both exhaust head pipes removed to allow transmission removal. Some head and Y-pipes can be very difficult to remove, so plan to soak the exhaust manifold bolts with rust penetrant prior to removal.

Before attempting to lower the transmission, make sure it is effectively strapped to the transmission jack. Most transmission jacks come with a chain or strap for this purpose. If using a floor jack, or pure muscle from a couple of strong helpers, take care to avoid being injured. A complete TH400 transmission weighs nearly 150 pounds, which is enough to cause some pretty serious damage if it falls on you when lowering it to the ground!

19 Use Safety Strap

A safety strap or chain is used to keep the transmission from sliding off the transmission jack during removal and installation.

20 Modify Floor Jack

A floor jack can be modified for use as a transmission jack by welding a flat metal plate to a piece of iron pipe. The iron pipe must fit snugly into the recessed hole in the top of the floor jack. Holes can be drilled in the metal plate, and bolts used to attach the plate to a transfer case mount. The modified floor jack in the picture has been successfully used to pull four-wheel-drive transfer cases and transmissions as one unit.

Most transmission jacks also have adjusters to tilt the plate in both directions. These are used to help keep the transmission in alignment with the engine during installation, but also to help keep it in alignment during removal. It is important to pull the transmission straight back off the engine dowel pins while maintaining correct alignment with the engine.

Carefully lower the transmission to the ground while paying close attention to transmission cooling lines, speedometer cable, shift linkage, or any other items that could get damaged during this procedure. Some applications require that the filler tube and dipstick be removed prior to lowering the transmission. With some applications, it's nearly impossible to install the filler tube after the transmission is in place, so it has to stay in the transmission during removal and installation

After the transmission is safely lowered from the vehicle, completely drain the unit. Do this by placing it in a transmission holding fixture with the tail housing down, or leaning it against a shop wall with the tail housing in a drain pan.

21 Utilize Transmission Jack Adjusters

Transmission jacks have threaded adjusters so the installer can get perfect alignment with the engine when installing the transmission. This is an excellent feature.

TH400 Disassembly

General Motors introduced the TH400 transmission in 1964. It was first used in Buick and Cadillac applications, and was quickly adapted to other models. It became the workhorse of General Motors' automatic transmissions, used in truck applications and anywhere else there was a need for a strong unit to handle high-torque engines, heavy vehicles, and towing applications.

TH400s make an excellent high-performance automatic transmission, and have been used for decades for all sorts of racing applications. The factory conservatively rated their torque input capacity at 450 ft-lbs, but they can handle much more. The performance aftermarket stepped up with heavy-duty parts to make them even better.

The gear ratio for low gear is 2.48:1, 1.48:1 for second, and 1:1 in third gear. A fully assembled unit weighs nearly 135 pounds, so it's no lightweight. Although often taunted as a bit heavy and power robbing, the TH400 makes up for it with its great strength and long-term durability in heavy-duty service.

The TH400 is also a very simple unit, and is relatively easy to overhaul. It does not require a lot of special tools, or special procedures. Most hobbyists have great success building their own units at home. The aftermarket offers plenty of heavy-duty and high-performance replacement parts for just about everything inside a TH400. Even so, the TH400 doesn't need a lot of help or special aftermarket parts to be a reliable and durable transmission. (I address potential weak points and recommended upgrade areas in the overhaul procedure section.)

Here I recommend and outline the replacement of all of the bushings during the rebuilding process. I am also replacing all of the thrust washers. Although not mandatory for all builds, doing so produces a unit that has firmer shifting, reduced gear noise, and ultimately lasts longer once placed into service.

Rebuilding an automatic transmission may be intimidating to some, but don't let its complexity overwhelm you. Stay focused, complete one step at a time, and eventually you'll have completed the task.

Remove External Components

1 Remove Modulator

Remove the bolt for the transmission modulator retainer.

Remove the modulator and the modulator valve.

A strong magnet usually pulls the valve from the case. If it doesn't want to come out, you can use a small flat-tip screwdriver to help it out of the case after the valve body is removed.

2 Remove Governor

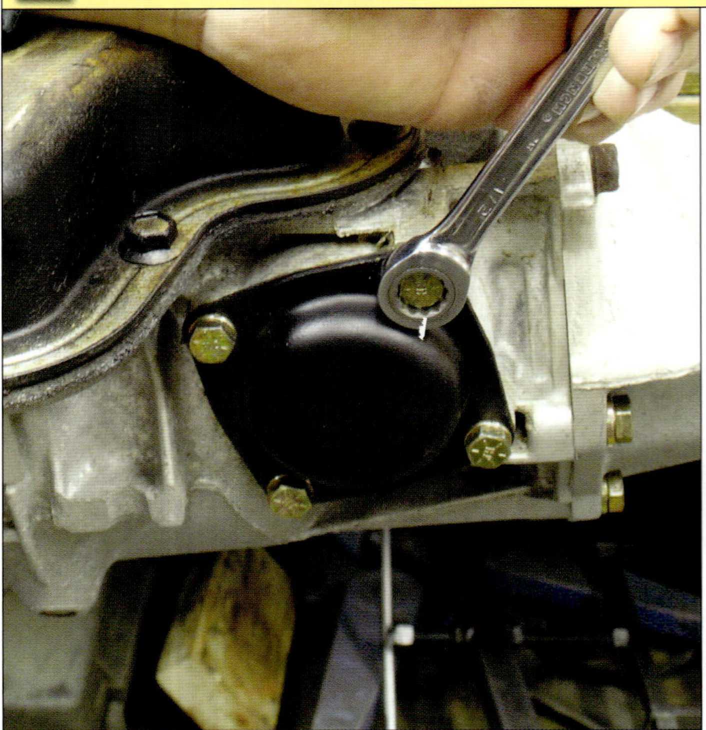 

Remove the governor cover bolts, governor cover, and the governor. Inspect the plastic gear closely for nicks, wear, and other damage.

3 Remove Tail Housing

Remove six tail housing retaining bolts and slide the tail housing off the transmission.

4 Remove Oil Pan

Remove the oil pan bolts, and carefully lift the oil pan up and off the transmission. It makes a good place in which to put the valve body and other associated parts removed under the pan until they can be cleaned.

Remove Valve Body and Governor Filter

1 Remove Oil Filter

Remove the special bolt that holds the oil filter in place. TH400s used several types of filters; the one shown is the most common. Some heavy-duty truck applications use a deeper pan and extension for the filter. Very early units used a side-mounted filter. Lift the filter off the extension tube, taking care not to break or crack the plastic tube.

2 Remove Filter Extension and O-Ring

Remove the plastic filter extension tube from the case. Remove the O-ring from the case. It's a common practice for builders to install two O-rings in the case, so there may be more than one; just make sure to get them out of the case at this time.

3 Remove Valve Body

Remove the valve body bolts and remove the valve body from the case. Three of the bolts are 1/4-inch x 20 threads (shown), requiring a 7/16-inch socket. The rest of the bolts are 5/16-inch x 18 threads and require a 1/2-inch socket to remove them. They are similar to the oil pump bolts but not interchangeable, so keep them separate.

Oil Tube Leads

The valve body has two oil tubes leading to the back of the case for governor oil supply and return. These tubes are easily bent or damaged. Make sure to gently pry them up at the same time the valve body is being removed. Once clear of the case, they slide right out of the valve body.

4 Remove Governor Filter

Locate and remove the governor filter. It should be found under the governor tube at the back of the case. It may be missing; many builders do not install one during transmission rebuilding.

Remove Downshift Solenoid and Separator Plate

1 Remove Downshift Solenoid Bolts

Remove the two 1/4-inch x 20 bolts that attach the electric downshift solenoid to the case. There were several different styles used. Some have a seal as part of the solenoid, while others require a separate seal.

2 Remove Downshift Solenoid

Locate and remove the downshift solenoid wire from the terminal, and remove the solenoid from the transmission.

3 Remove Separator Plate

The separator plate and separator plate gaskets can now be removed.

Remove Front Band and Low Band Apply Servos

1 Remove Front Band Apply Servo

Lift the front band apply servo from the case. Be careful not to let the pin slide out of the piston and lose the clip that retains the spring retainer. It is best to leave this assembly intact so none of the parts get lost or re-arranged during cleaning.

2 Remove Check Balls

Remove the six check balls from the case, noting their locations.

3 Remove Low Band Servo Cover

Remove the six bolts that attach the low band servo cover to the case. Lift off the servo cover, then remove the apply piston. Note that inside the apply piston is the 1-2 accumulator and spring; remove them as well.

4 Remove Parking Pawl Guide

Remove the parking-pawl-guide retaining bolts and lift the guide out of the case.

Remove Oil Pump, Manual Shaft and Speedometer Gear Housing

1 Remove Center Support Bolt

Using a 12-point 3/8-inch socket, remove the center support retaining bolt. This is a special hollow bolt; oil must flow through it to the intermediate clutch. Put it in a safe place until reassembly of the unit.

Prior to removing the oil pump, it is a good idea to take an input shaft endplay reading. The factory used specific shims and selective thrust washers under the pump to set endplay. If you purchase a thrust washer kit, it usually comes with selective shims and thrust washers to set endplay during the rebuild. Endplay should be no more than .030 inch, and no less than .015 inch.

2 Remove Oil Pump

Remove the oil-pump-to-case bolts; there may be either 6 or 10 bolts, depending on the year of the transmission. Two of the pump bolt holes are threaded for 3/8-inch x 16 threads. Using a slide hammer, remove the oil pump from the case.

2 | Remove Oil Pump *continued*

3 | Loosen Nut

Loosen the nut on the end of the manual shaft.

4 Remove Manual Shaft

Remove the retaining pin for the manual shaft, and slide the manual shaft out of the case. You can now remove the transmission linkage and parking pawl shift rod from the case.

Smooth the Sharp Edges

It may be necessary to use a jeweler's file to remove any burrs or raised areas from the edges of the manual shaft, where it went through the linkage, in order to get it to slide out of the case.

5 Remove Speedometer Gear Housing

Remove the speedometer gear-housing retaining bolt and retainer. Pry the speedometer gear housing from the case. Note that I have installed the end of an old speedometer cable to pry on it to avoid damaging the threads.

Remove Forward Drum Clutch, Direct Drum and Band

1 Remove Forward Drum Clutch

Get a firm hold on the input shaft and pull the forward clutch drum from the case.

2 Remove Direct Drum

Remove the direct drum from the case. It can be difficult to get a good grip on it; the outer snap ring and clutch and steel plates can be removed if needed (left). You can use two hooked screwdrivers to remove it (right).

3 Inspect Band

The band can now be removed from the case. Check the condition of the band lining and the surface on the drum where it applies.

Remove Clutch Pack, Output Shaft and Low Band

1 Remove Clutch Pack

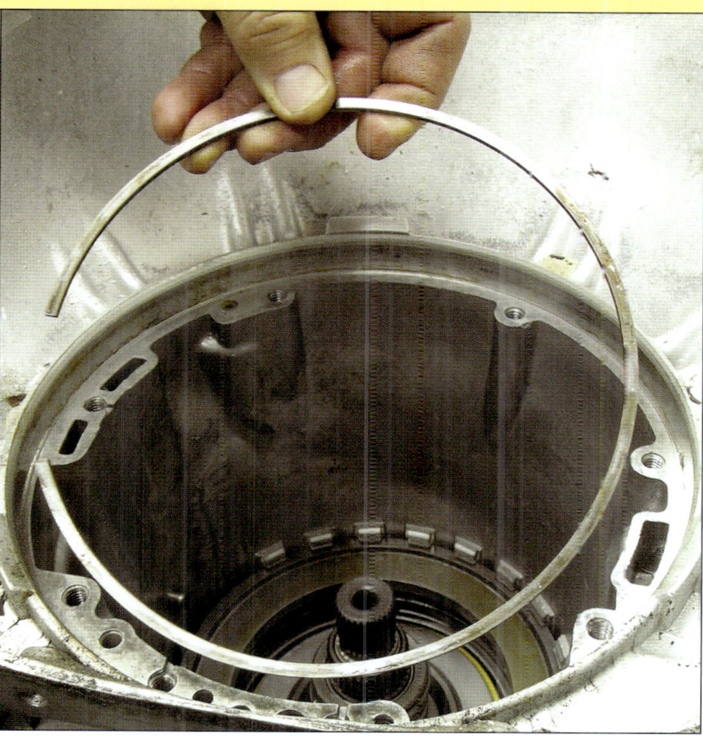

With a large screwdriver, remove the intermediate clutch retaining ring, backing plate, and clutch pack.

2 Remove Retaining Ring

Remove the center support retaining ring, center support, and backing spacer ring (not used on all models).

3 Remove Outer Shaft

Remove the outer shaft that splines into the sun gear.

4 Remove Output Shaft

Get a firm grip on the inner shaft, and lift the entire lower assembly/output shaft out of the case.

5 Remove Thrust Washer

Remove the thrust washer backing spacer (left) and low band (right) from the case.

Final Disassembly

1 Remove Solenoid Connector

Pull the downshift solenoid electric connector from the case. Use care not to damage the unit. Use a small screwdriver, if needed, to depress the retainers.

2 Remove Manual Shaft

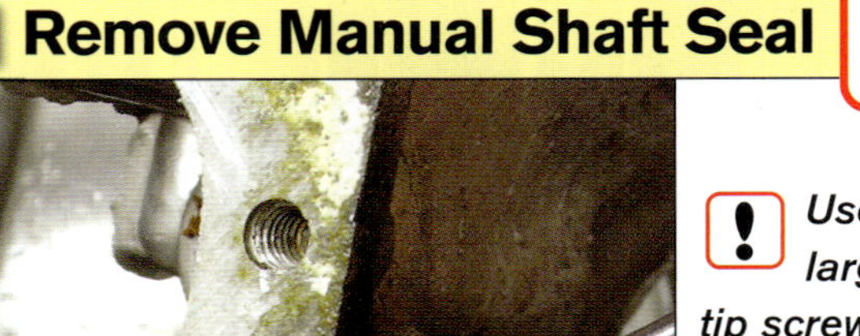

Remove manual shaft from the case.

3 Remove Manual Shaft Seal

Important!

Use a large flat-tip screwdriver to pry the manual shaft seal from the case. Use care not to damage the bore around the seal.

4 Remove Parking Pawl

Remove the parking pawl spring.

5 Strip and Clean Case

The case is now stripped down for cleaning. A power washer removes heavy deposits from the case. Soaking it with engine cleaner or a heavy-duty degreaser helps the process. The case can also be taken to a machine shop for professional cleaning. It must be clean and free of all dirt and debris prior to assembly.

TH400 ASSEMBLY

Now that the case is cleaned you are ready to start transmission assembly. The transmission is built from bottom to top, starting with the bushing for the output shaft, and moving up through the unit from one sub-assembly to another.

If an aftermarket shift kit, or transbrake is being installed, refer to the specific instructions supplied with your "kit" during transmission assembly. Special modifications are required, such as drilling a bleed hole in the direct drum, leaving out the center seal on the apply piston, etc.

Some sub-assemblies require special procedures and tools, such as the removal of the snap rings above the spring cage using a suitable spring compressor on the drum.

Install New Case Bushing, Solenoid Connector and Parking Pawl

1 Replace Rear Case Bushing

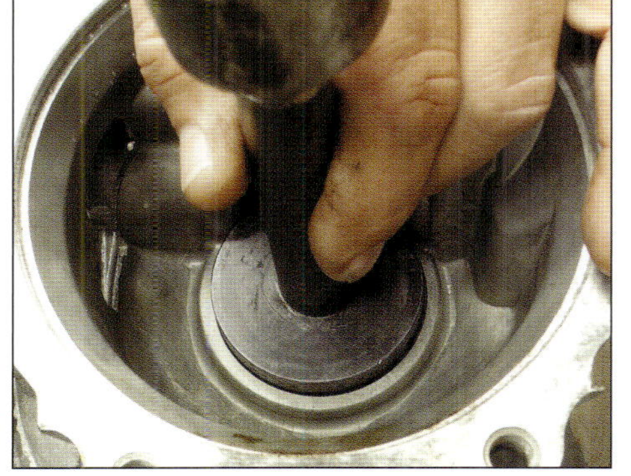

Once the case is cleaned and ready for assembly, drive out the rear case bushing for replacement. Using a suitable driver, apply a small amount of red Loctite to the new bushing and drive it in place.

2 Install Solenoid Connector

Install the solenoid connector into the case with a new O-ring. Use a small amount of transmission fluid or Transgel so it slides easily into place.

3 Install Manual Shift Seal

Install the manual shaft seal into the case with a driver until it is fully seated. Avoid collapsing the seal during installation, or it will leak when the transmission is placed in service.

4 Install Parking Pawl Spring

Hook the parking pawl spring around the parking pawl, then on the transmission case. This holds the parking pawl out of the way so the transmission output shaft and lower assembly parts can be installed.

Disassemble Output Shaft and Lower Planetary Assembly

1 Inspect Parts

 *The output shaft and lower planetary assemblies and related hardware are shown here. Remove the low planetary carrier from the output shaft carrier assembly to inspect the pinions, Torrington bearings, and replace the bushing.*

2 Remove Snap Ring

With a flat-tip screwdriver or other suitable tool, remove the snap ring to separate the lower planetary assembly from the output shaft assembly. This allows access to the two sets of Torrington bearings, and to replace the output shaft upper bushing.

3 Drive New Bushing

Note the depth of the old bushing and make sure to drive the new bushing to the same depth, and line up the lube oil hole at the same time. Use a sharp-angle cut punch to catch the edge of the old bushing and remove it. Use care to avoid damaging the material where the bushing rides.

Use fine-grit automotive sandpaper to remove any scratches, if present, before driving in the new bushing. There may be two different width measurements for these bushings; the wider bushing is the preferred replacement.

4 Install Bearings

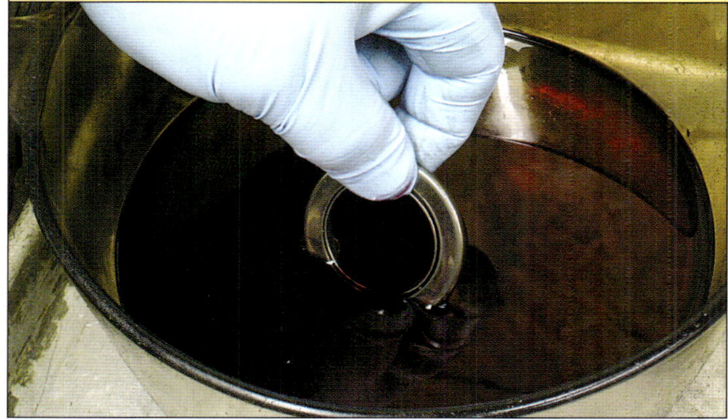

Lube both lower Torrington bearing sets with clean ATF before reassembly of the lower unit. Make sure they are installed correctly with the lips facing in the same direction and in the same location from which they were removed. (See Step 2 on page 59 for more detail.)

Reinstall Planetary Assembly

1 Install Planetary Carrier

Install the low planetary carrier, output shaft, and snap ring.

2 Install Thrust Washer

Install a new thrust washer, tangs down, and use a small amount of transgel to hold it in place.

3 Install New Bushing

Drive out the old bushing in the front planetary assembly. Apply a small amount of red Loctite to the new bushing and install, driving it flush with the upper edge of the bore.

4 Replace Roller Clutch Assembly

Replace the low roller clutch assembly. For spring assemblies, see sidebar. Lubricate the one-way clutch assembly with clean ATF.

1964–1966 (early) Sprag Assembly

1 Note Assembly Type

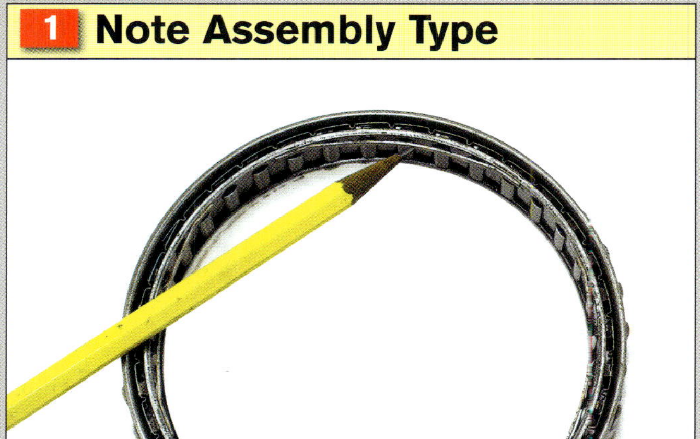

Very early units used a sprag assembly instead of a roller clutch assembly. It has a built-in retainer because the assembly process is slightly different than units that used a roller clutch.

2 Smooth Race

The sprag-type clutch requires a planetary assembly with a smooth outer race.

3 Install Sprag Assembly

The sprag assembly must be installed on the bottom of the center support assembly, and then lowered into the case.

4 Install Center Support

As the center support is lowered into position, it stops when the sprag rests on the smooth race inside the low planetary assembly. When the center support stops, reach down and turn the output shaft. This allows the sprag assembly to mesh with the smooth inner race and the center support to lower into the case so you can install the snap ring. The center support must be fully seated to provide enough room to install the snap ring.

5 Assemble Output Shaft

Lubricate the planetary pinions, install the silencer ring, and lower the upper planetary assembly in place. Turn the unit in both directions until it drops into position and is fully seated.

6 Install O-Ring

If used, install a new O-ring onto the output shaft.

7 Install Thrust Washer

Install a new thrust washer on the lower end of the output shaft assembly, and install the speedometer drive gear if removed previously.

8 Install Thrust Plate

Install the case thrust plate/spacer. Use a small amount of transgel on the back side to retain it to the case.

Install Low Band and Output Assembly

1 Install Low Band

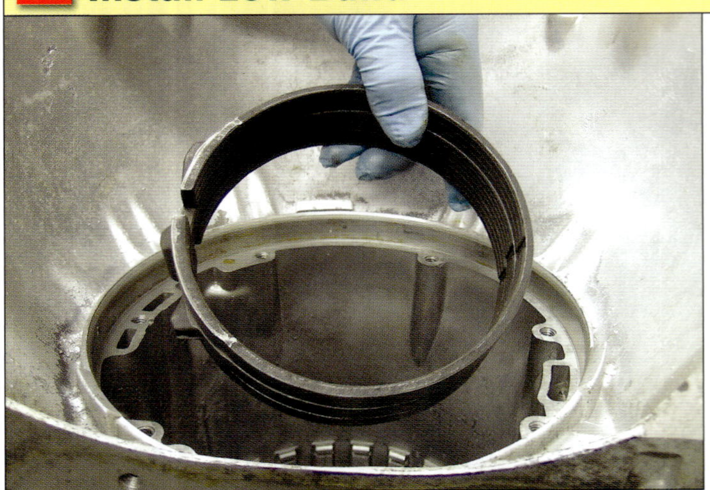 

Install the low band into the case. Lubricate the band lining with clean ATF.

2 Install Output Assembly into Case

The entire output assembly can now be lowered into the case. Do not dislodge the low band during this procedure. Once it is fully into position, you can install the sun gear (above left). Note the orientation of the drum with the low band. Also note that the sun gear is installed with the inner chamfered edge down toward the back of the case. The flat side faces up toward the center support and upper Torrington bearing.

3 Install Lower Spacer (if applicable)

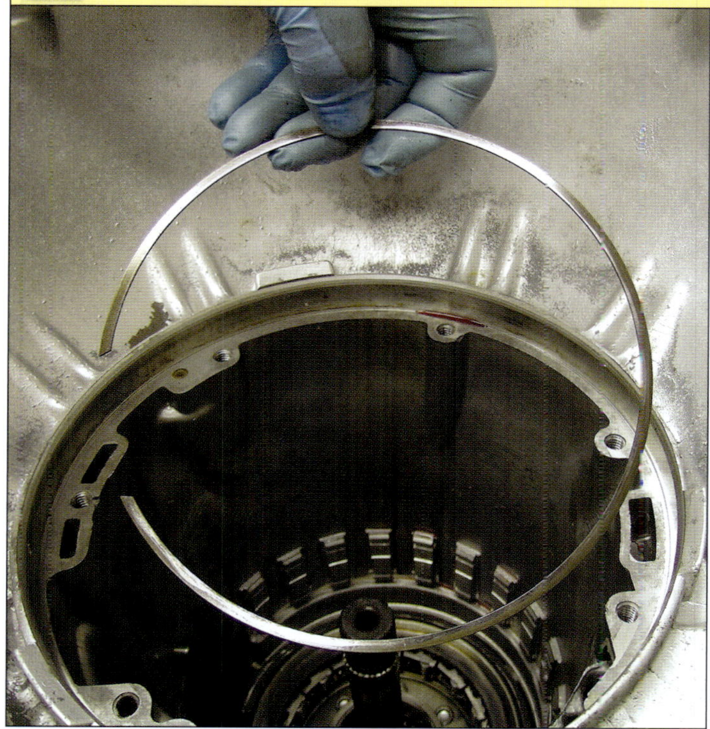

If used, install the lower spacer for the center support assembly. Early models did not use this spacer, so use care when switching out parts between early and late cases. There should be no space between the center support snap ring and the center support when the correct parts are used. If the snap ring does not go into place, or there is a gap between the snap ring and the center support, then the wrong center support has been installed.

Rebuild Center Support

1 Remove Thrust Washer

Remove the thrust washer from the bottom of the center support, and the sealing rings from the grooves.

2 Remove Apply Piston

Remove the apply piston retaining ring, springs, spring carrier, and apply piston from the center support.

3 Remove Center Support Bushing

Drive out the center support bushing using a suitable driver. Make sure to provide sufficient support for the center support during this procedure.

Critical Inspection

4 Inspect and Verify Parts

All of the parts for the center support are shown here. Some models use an aluminum apply piston instead of a steel one. Replace the thrust washer, bushing, apply piston seals, and sealing rings. Inspect the Torrington bearing assembly, and replace it if needed.

Reinstall Center Support

1 Install New Bushing

Drive a new bushing into the center support assembly, taking care to align the oil holes and driving it flush. Make sure to support the unit on a strong wooden block to avoid damaging it during this procedure.

2 | Install Lip Seals

Install two new lip seals on the intermediate clutch apply piston. Install the apply piston into the center support, using plenty of lubricant, clean ATF or transgel, and a lip seal installation tool. Once it is fully seated, install the spring guide, three springs, spring retainer, and retaining snap ring.

3 Install Sealing Rings

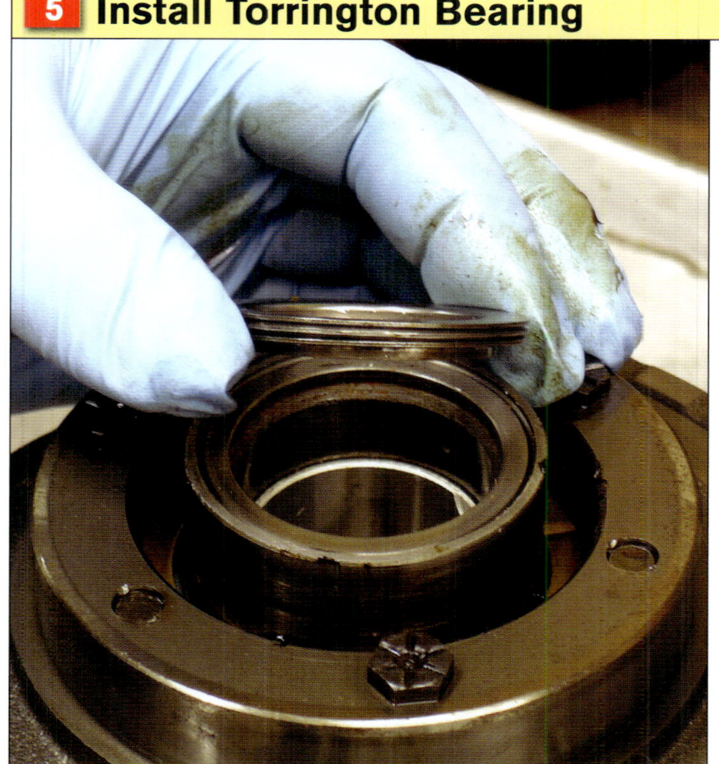

Install four sealing rings into each ring groove. If using the factory-type cast-iron hooked rings, make sure to hook the ends together once the rings are in the ring grooves.

4 Install Thrust Washer

Install a new thrust washer in place on the underside of the center support, and retain with transgel. Most replacement thrust washers are metal to replace the factory plastic thrust washer.

5 Install Torrington Bearing

Install the three-piece Torrington bearing assembly to the bottom of the center support. Use a generous amount of transgel to hold all three parts together and in place while the center support is lowered into the case.

6 Seat Retaining Ring

Lower the center support into the case, making sure the oil holes are aligned correctly. It may be necessary to rotate the output shaft to allow the unit to fully seat. Once fully seated, install the retaining ring with the open section toward the area of the case where no case lugs are present. The retaining ring for the center support is tapered and not square cut. Make sure to install the retaining ring with the flat portion against the center support. Using a large screwdriver, fully seat the retaining ring into the grooves in the transmission case.

7 Replace Bushings

Remove and install new bushings into each end of the intermediate shaft, and install it into the transmission until it is fully engaged with the sun gear. It may help to turn the output and input shaft during installation to get the intermediate shaft to fully seat into position.

Rebuild and Install Clutch Pack and Intermediate Band

1 Install Intermediate Clutch Pack

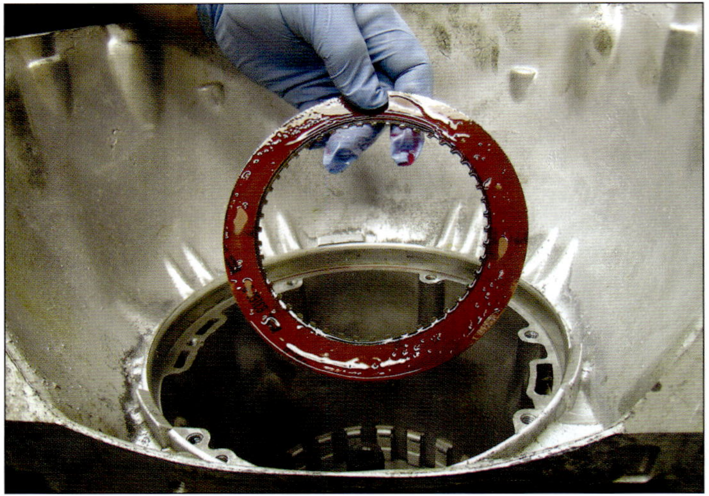 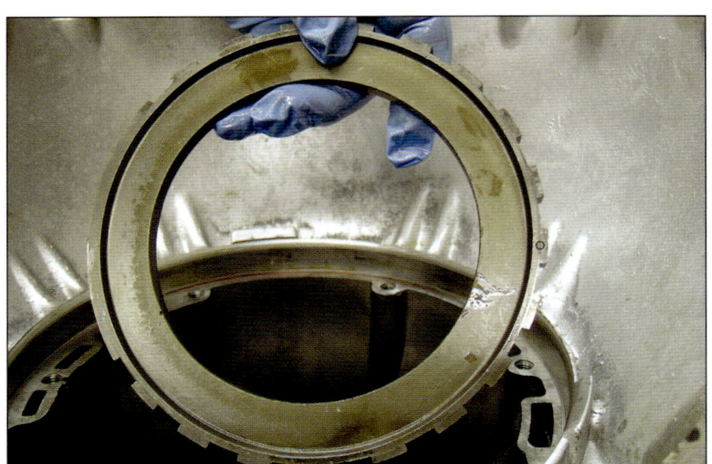

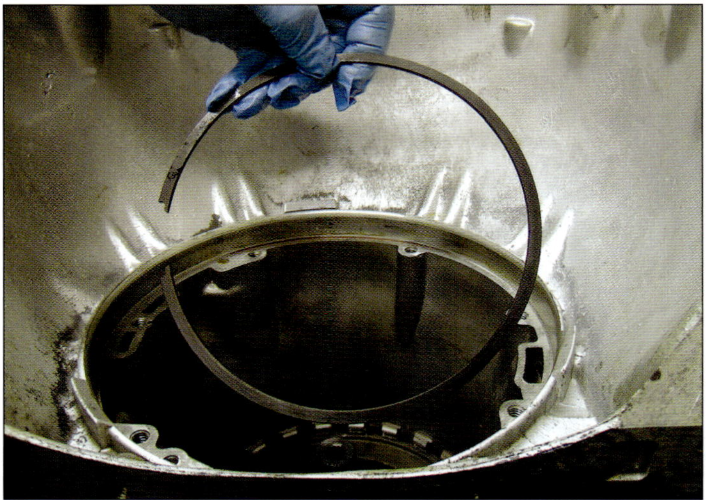 

Install the intermediate clutch pack. Make sure to soak the frictions for several minutes in clean ATF prior to installation. Start with a steel plate, then alternate steel/friction, steel/friction, ending with the thick steel backing plate. Install the retaining snap ring, with the open end facing the portion of the case where no case lugs are present.

Waved vs Smooth Plates

The factory used several different clutch pack arrangements for the intermediate clutch pack in the case. The most common has smooth, "waved" frictions and thick, flat, steel plates. Yes, the smooth factory frictions have a very sight "wave" to them. The first steel plate may be "waved" as well. Some units were also set up with flat "waffled" frictions.

Most rebuild kits contain the thick, flat steel plates and smooth waved frictions. They usually have one waved steel plate as well.

As a general rule, it is best to duplicate what was removed from the unit for a stock rebuild. The waved-steel plate is fine, but can be replaced with one thick steel plate to shorten the shift. When combined with the factory-type smooth "waved" frictions this makes for the easiest replacement setup, even for high-performance rebuilds.

Another popular option for high-performance rebuilds is to install the clutch pack setup used in the later 4L80E units. This contains four frictions, instead of three, and thinner steel plates.

The important thing to remember here is to make sure that you have duplicated the total thickness of the clutch pack that was removed from your unit, if you deviate from that setup with different steels and/or frictions.

2 Install Retaining Bolt

Install and tighten the retaining bolt for the center support. Using a large Phillips screwdriver or other suitable tool, apply torque to the center support while tightening the retaining bolt.

3 Test Clutch Pack

After the center support and intermediate clutch pack installation is complete, air test the clutch pack.

4 Install Band

Install the intermediate band into the case.

5 Sprag Upgrade

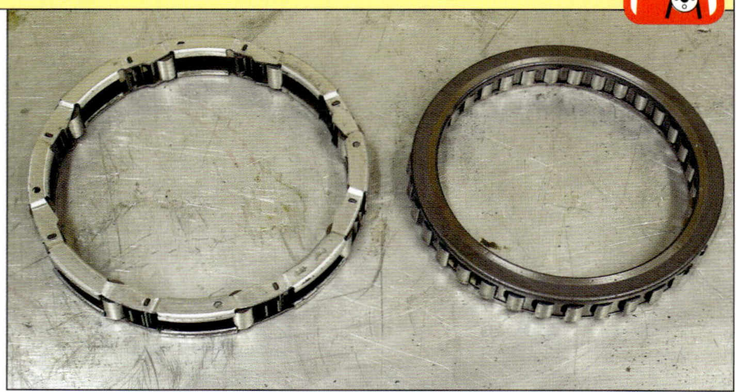

Two types of direct drums were used in the TH400. The notched drum (left) uses a roller-type clutch, while the smooth drum (right) uses a sprag-type clutch.

There is a 34-tooth upgrade (right) for the smooth drum, and it is highly recommended for any TH400 transmission to be used in a high-performance application.

6 Note Piston

The direct apply piston may be made of cast aluminum (left), or steel (right).

7 Upgrade Retaining Ring

The factory retaining ring for the intermediate one-way clutch is shown on the left. The later 4L80E transmission uses a spiral locking ring at the same location. It is best to use the upgraded retaining ring for all high-performance TH400 transmission rebuilds.

8 Note Components

Here are the basic components of the direct drum. Several different clutch pack configurations were used for various models, so pay close attention to what was removed from the original drum, so the total clutch pack thickness can be duplicated during the rebuild.

9 | Install Intermediate Clutch

Install the intermediate clutch. Start with a lower race, sprag, outer sprag race, upper race, retainer, and snap ring. If a sprag-type clutch is being installed, the outer race should turn freely in the clockwise direction, but lock solidly in the counterclockwise direction. If a 4L80E spiral lock ring is being installed, make sure to snap it into position while winding it into the ring groove. This may take considerable effort, so be careful to not damage the retaining ring during installation.

10 Install Drum Seal

Install the back-up seal onto the direct drum, with the lip facing up.

11 Use Check Ball

All direct drums use a check ball in either the drum (above left) or in the apply piston (above right), but not in both locations. If a new apply piston or drum is being installed, make sure that one of the two components uses a check ball. If both components end up with a check ball, one ball can be eliminated by removing it and installing a set screw to block oil flow (shown at left).

12 Install Clutch Apply Seals

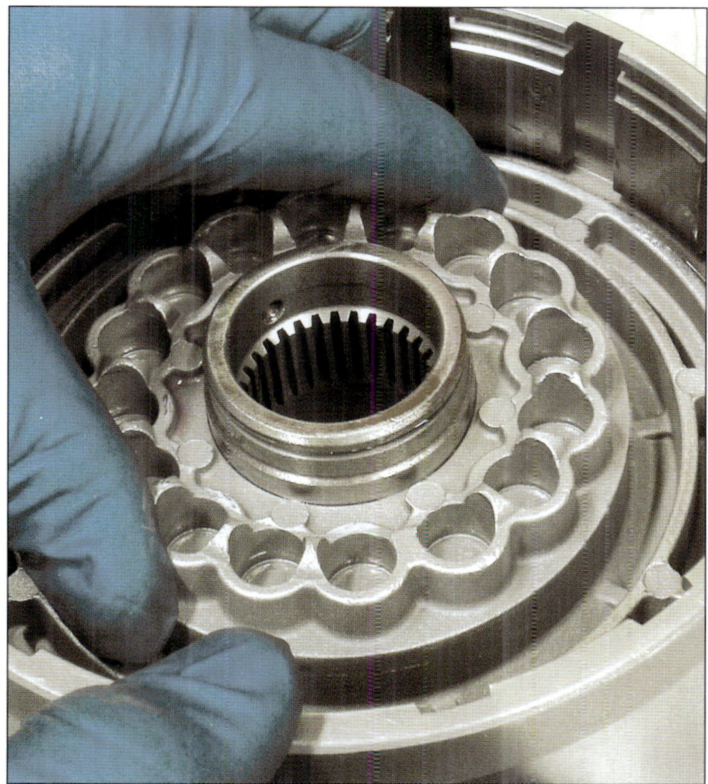

Install both direct clutch apply seals, with the lips facing down. Using plenty of lubricant, and a seal installation tool, install the apply piston into the drum until it is fully seated.

13 Install Piston Return Springs

Install the piston return springs and spring retainer, and compress the retainer enough to install the retaining snap ring. Make sure the snap ring ends are between the raised areas of the retainer.

Rebuild Direct Clutch Pack and Install Direct Drum

1 Build Clutch Pack

You are now ready to build your direct clutch pack. Note that several different arrangements were used for different models. Steel plates could be in several different thicknesses, and as many as six friction plates could have been used in the direct drum. Some drums also used a waved apply plate, which may have been directly on a friction, or against a steel plate and then a friction. Since many possibilities exist, you must ensure that the clutch pack has sufficient clearance, and uses the best combination of components for the application. The most common setup (shown) is to use five .090-inch-thick steel plates and five stock frictions. A waved plate can be substituted for the first thick steel plate, but this is not recommended for high-performance applications.

TECH TIP

Building a High-Capacity Direct Drum

You have the option to use thinner .078-inch factory steels, or even .060-inch aftermarket steels, and .060- or .078-inch frictions to build a high-capacity direct drum. A shorter apply piston can also be used, or a factory aluminum apply piston turned down in a lathe, to accommodate more frictions and steel plates, and/or to establish sufficient clutch pack clearance.

You must consider that even though is it quite easy to build a high-capacity direct clutch pack, using thinner components (specifically steel plates) reduces their ability to take heat and maintain their shape in a high-performance application. In any case, I recommended using at least five frictions in the direct clutch pack and the thicker .090-inch steels, if possible, with the combination(s) of parts available to you.

2 | Install Steels and Frictions

Install the direct clutch steels and frictions, starting with a steel plate, and alternating friction/steel, friction/steel, ending with the thick factory backing plate. Install the snap ring and check clutch pack clearance. It should be at least .015 inch (.030 to .060 inch is preferred), especially for high-performance applications.

3 | Install Direct Drum into Case

You can use an old outer sprag race to align the intermediate clutch plates in order to facilitate the installation of the direct drum into the case. The drum does not drop easily into position, so take the time to align and center the clutch plates into the case.

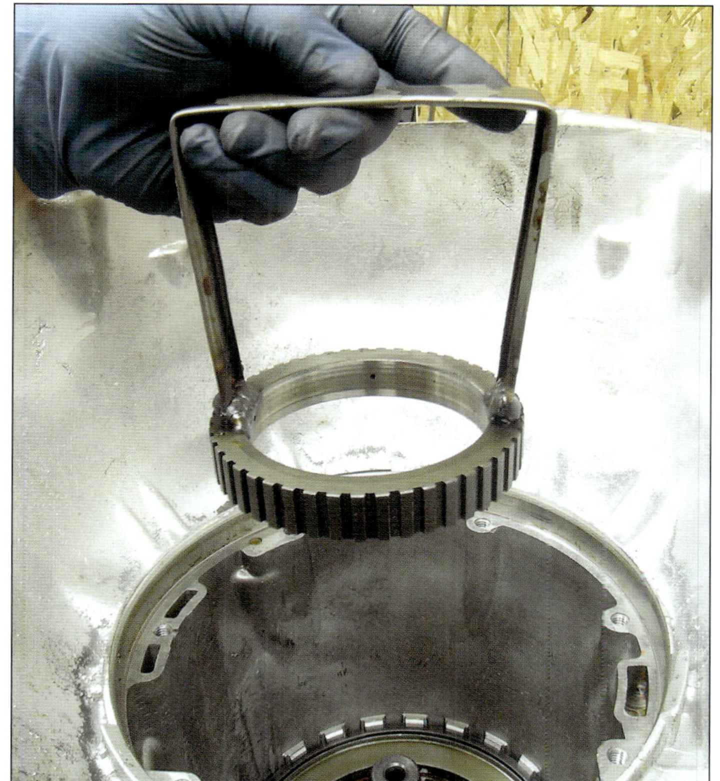

I welded a handle onto a old sprag outer race to use as an alignment tool for the intermediate frictions. It works well.

You can also apply air through the hole in the case to the intermediate piston, to "bounce" the clutch pack and help the direct drum drop fully into position.

1 Disassemble Forward Drum

Rebuilding the forward drum is similar to rebuilding the direct drum. Remove the snap ring, and then the backing plate, inner splined hub, and clutch pack. Note that thrust washers are used on both sides of the inner splined hub.

2 Remove Springs

Using a suitable spring compressor, remove the spring cage snap ring, spring retainer, springs, apply piston, and inner seal.

3 Note Check Balls

Here are the forward clutch and related components. As with the direct clutch, the apply piston may be made of steel or aluminum, and the check ball could be found in either the drum or the apply piston, but not in both locations.

4 Replace Lip Seals

Remove and replace the lip seals on the apply piston (with the lip facing down), then the back-up seal on the drum, with the lip facing up. Make sure to clean out and blow compressed air through the lube oil holes on the input shaft.

5 | Install Piston and Springs

Lubricate the back-up seal, and apply piston inner and outer seals.

Using a seal installation tool, install the apply piston into the drum, making sure not to cut or fold over the lip seals during installation.

Make sure the apply piston is fully seated, and turns freely in the drum.

Install the piston return springs. Recesses are provided for them if an aluminum apply piston is used.

5 Install Piston and Springs *continued*

Install the spring retainer.

Compress the retainer with a suitable spring compressor and install the snap ring, making sure the open ends are not aligned with any of the raised areas of the spring retainer.

If using a steel apply piston, as shown here, install the spacer ring on top of the apply piston.

6 | Install Plates

Install the steel and friction plates, followed by the backing plate and snap ring. I recommend a waved apply plate for all stock builds; it cushions the apply when the transmission is placed in gear. Depending on the efficiency of the torque converter and idle speed, the apply pressure that you feel can be pretty harsh if the waved apply plate is left out of this drum. Check the endplay of the clutch pack. It needs some play, but not nearly as much as the direct drum, .010 to .020 inch is recommended.

Building a High-Capacity Forward Drum

As with the direct drum, several different clutch pack setups were used for various models. Some have a waved apply plate directly on the first friction, which is fine. Some have the waved plate against a flat steel plate, and then the first friction. Most forward drums use the thinner .078-inch steel plates instead of the .090-inch. By leaving out the waved plate, and experimenting with different thicknesses of steels and frictions, a high-capacity forward drum is easy to build. It also helps if you have access to a lathe to remove some material from the apply piston as well, in order to build a custom higher-capacity drum. A high-performance forward drum should have at least five frictions, regardless of the combination of steels and apply piston thicknesses used to build it.

7 Complete Forward Drum Assembly

Complete the assembly of the forward drum once the clutch pack setup is determined and checked for sufficient endplay. Install the inner splined drum, making sure the lower thrust washer is in place and held there with a small amount of transgel. Install the backing plate and snap ring, and fully seat the snap ring with a large flat-tip screwdriver. Be sure the upper thrust washer is in place, and retained with transgel. The drum is now ready to install in the case.

8 Install Forward Drum

Install the forward drum into the case. Make sure it is fully seated. The best procedure to use is to wear a leather glove, and twist the input shaft back and forth to help align the frictions in the direct drum with the splines of the forward drum. It drops down one friction at a time until fully seated; this may take several minutes.

Rebuild and Install Oil Pump Assembly

1 Rebuild and Install Oil Pump

The oil pump is the last component to rebuild. Flip it over and remove the five bolts that hold the two halves together. The upper section drops down, allowing the two halves to be separated.

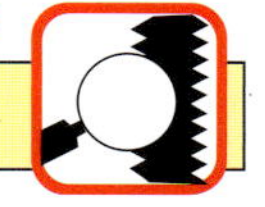

Critical Inspection

2 Inspect Pump Gears

Inspect the pump gears closely where they engage with the torque converter. This is a common area to find wear or damage.

3 Remove Seals and Bushing

Remove the outer pump sealing ring.

Remove the front seal.

Remove the bushing.

Use a sharp-angled punch to catch the edge of the seal. When driving out the torque converter bushing, do not to damage the area of the housing where it resides.

4 Remove Sealing Rings and Thrust Washer

Remove both hooked iron sealing rings and the front thrust washer. The front washer is selective, with several thicknesses available to set input shaft endplay during the rebuild.

5 Remove Snap Ring

Remove the snap ring that retains the pressure regulator valve, spring, and related components. (The correct orientation of the parts is shown here.) Some valves also have "horseshoe" shims behind the spring retainer. Make sure to use them unless the pressure regulator spring is being replaced with one from a shift kit, and the instructions recommend to discard any "horseshoe" shims found behind the retainer.

Gently clamp the rear pump half in a soft-jawed vise. Compress the pressure regulator spring slightly and remove the retaining snap ring with a pair of snap ring pliers.

6 Adjust Feed Hole

I recommend reducing the size of the converter's feed hole for high-performance transmission builds. This helps to ensure that fluid pressure used to fill the converter doesn't push it unnecessarily hard against the engine's thrust bearing.

Tap the converter feed hole about five to six turns with a tapered 5/16-inch x18 tap. Install a brass 5/16-inch set screw.

Drill the set screw to .150 to .160 inch.

This still allows plenty of fluid for effective converter filling, and also reduces any tendency for excessive pressure to blow out the front seal.

7 | Remove Inner Bushings

Using a sharp-angle cut punch, remove both pump inner bushings by driving them down on one side, and then prying them out.

Make sure to measure the depth of each bushing before driving them out, so the new bushings can be driven into the correct position.

8 Install New Bushings

Install new pump bushings, taking care to drive them to the correct depth below flush. Use a drop of red Loctite on them so they stay in place once pressed in service. It is a good idea at this point to test fit them by sliding the lower housing down over the forward drum input shaft. They may be a bit snug, but the shaft should turn freely inside the bushings.

9 Install Torque Converter Bushing

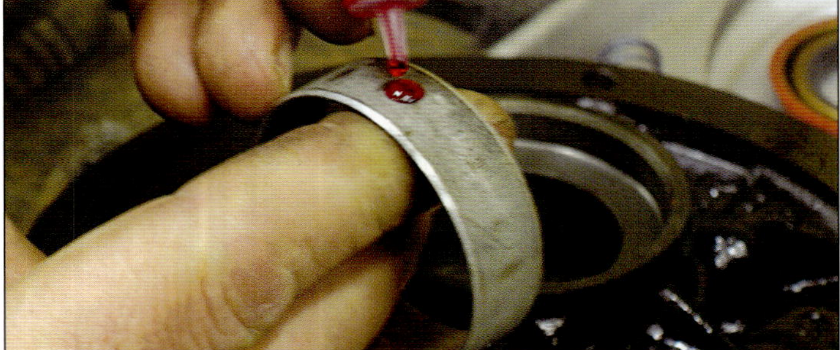

Apply a drop of red Loctite to the new torque converter bushing and drive it in place. Drive from the front and seat the bushing flush with the upper edge of the bore.

10 Install Front Seal

Apply a drop of Loctite to the front seal, and drive it into the pump housing until fully seated. Make sure that the seal's inner spring doesn't come out during this procedure.

11 Install Oil Pump Gears

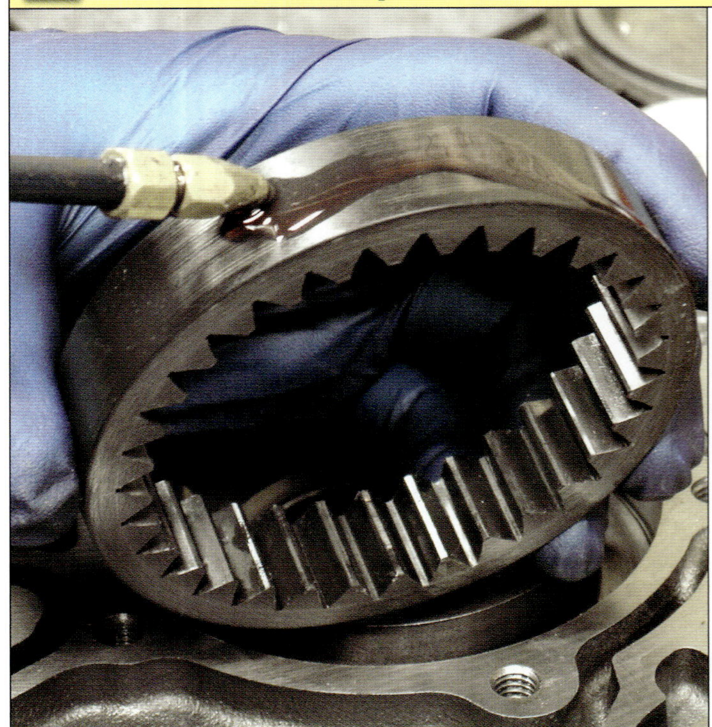

Lubricate and install the oil pump gears in the same orientation as they were removed.

The inner gear drive always faces away from the torque converter. If the inner gear is installed backward, it destroys the oil pump when placed in service.

Always check the pump clearance to the flush surface of the pump half. The pump gears should turn smoothly and freely within the housing.

12 Install Front Washer

Measure the thickness of the front thrust washer. Use a new one of the same thickness if your rebuild kit came with an assortment. If not, reuse the stock thrust washer, provided it is in good condition.

13 Install New Sealing Rings

Install two new factory sealing rings, making sure the ends are hooked together and the rings move freely in their respective grooves.

14 Align Pump Halves

Place both pump halves together and hand tighten the bolts. Use a large hose clamp to align the pump halves while ensuring good alignment of the bolt and oil holes. Tighten the bolts once the pump halves are correctly aligned. The bolts should be torqued to 18 to 20 ft-lbs.

15 Install Rubber Seal

Install the large outer rubber seal into the groove in the pump. Take care not to twist the seal in the groove during installation.

16 Install Pressure Regulating Valve

Lubricate the pressure regulator valve and install into the corresponding bore in the oil pump.

The spring retainer goes against the valve, with any "horseshoe"-style shims that were removed.

Install the shims against the valve, then the retainer, followed by the pressure regulator spring.

Install the inner valve and retain it with some transgel or petroleum jelly. Install the pressure regulator spring (then the outer sleeve and valve).

16 Install Pressure Regulating Valve *continued*

Compress the assembly and install the snap ring, with the flat side facing up. It helps to gently clamp the pump in a soft-jawed vise during this procedure. Push down several times on the pressure-regulating valve assembly to make sure the snap ring is fully seated.

17 Lubricate Seal

Lubricate the outer pump seal with clean ATF.

18 Install Pump Gasket

Install the pump-to-case gasket and a couple of guide studs to help align the pump while it is lowered into position.

19 Install Manual Shaft

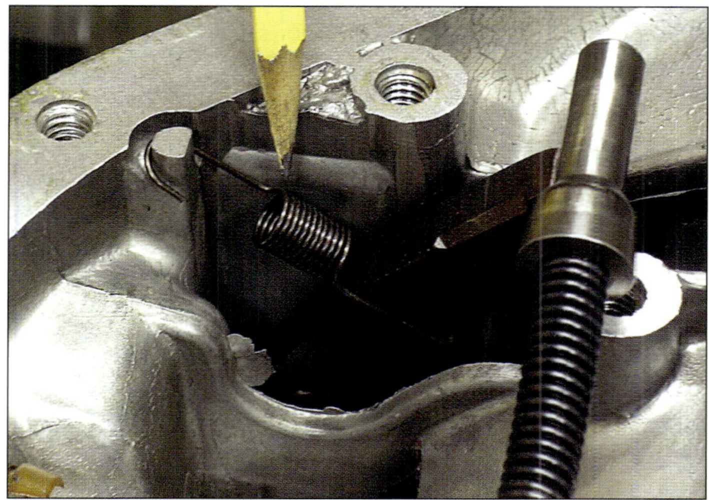

It is best to install the manual shaft and related hardware before installing the oil pump. This makes it much easier to install the retaining pin for the manual shaft. The pin can still be installed with the pump in place, but it has to be bent and then straightened again. Make sure that the parking pawl retainer is installed correctly. It is possible to install it backward, and then the vehicle will not hold when the selector is placed in the Park position.

20 Install Oil Pump

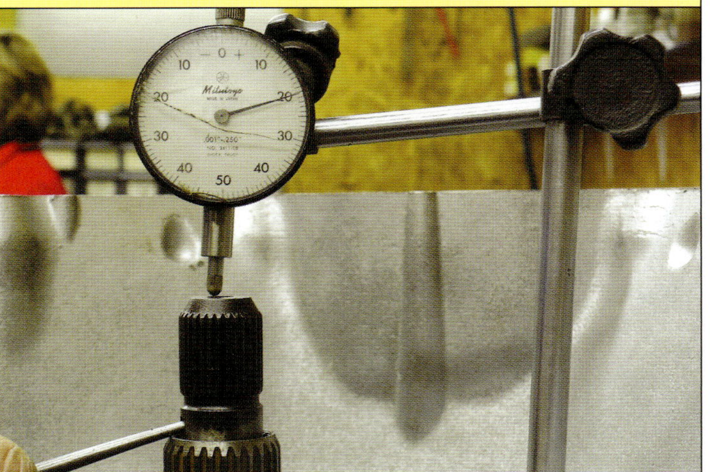

Lower the oil pump into place, and install the bolts with new seals from the rebuild kit.

Once the pump is tightened, check the input shaft end-play. A small awl or screwdriver can be used to pry upward on the input shaft. Shaft endplay should be .015 to .030 inch. Use selective shims under the Torrington bearing, or selective washers and shims to correct the pump endplay to within specifications. Keeping endplay to a minimum, and installing a full set of bushings reduces up/down and end-to-end movement of the transmission's internals. This improves sealing at the sealing rings, reduces gear noise, improves transmission shift function, and increases transmission life.

Check Clutch Packs and Install Accumulator Piston

1 Check Clutch Packs

Once the oil pump is installed and the bolts tightened, air test the clutch packs for correct operation. The intermediate clutch is applied by forcing compressed air through the center support retaining bolt. The holes on either side of the bolt are for the direct drum (also used for reverse), and the hole in the case is used to check the forward clutch for proper application. This final air check ensures that you haven't broken any sealing rings when the parts were stacked in the case. A distinct "clunk" should be heard when applying the clutch packs, with minimal air leakage at the same time. If any excessive air leakage is noticed, the transmission should be taken back apart to investigate the cause.

Air-Pressure Check

When applying air to the holes left and right of the center-support retaining bolt, some air leakage through the opposite hole is normal, as the direct drum is also used for reverse. To get a positive seal, a small shop rag can be stuffed into the opposing hole when the air pressure check is performed.

2 Install New Sealing Ring

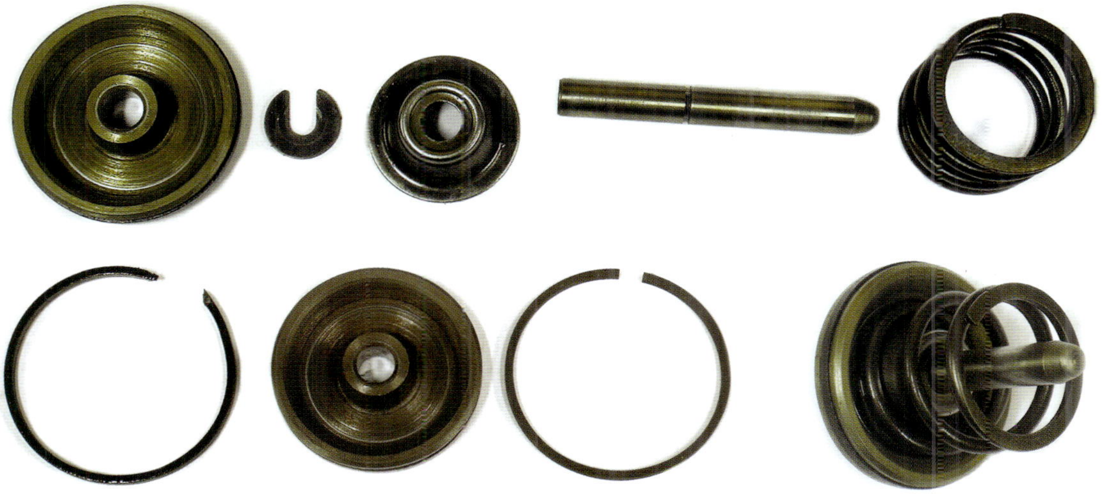

Install a new sealing ring on the front band servo piston. Some factory rings are made of Teflon. Most rebuild kits contain an iron ring to replace the Teflon ring. Make sure that the clip is installed above the spring retainer and stays in place as the parts are slipped together. Without the clip in place, the servo piston is not able to push the pin and the band does not work when the transmission is placed in service. Gently install the entire assembly into the case and push on the piston to verify band function.

3 Note Inner Spring

The low band apply servo also contains an accumulator. The inner spring does double duty here; it pushes off the band and operates the accumulator piston.

4 Install New Servo Seal

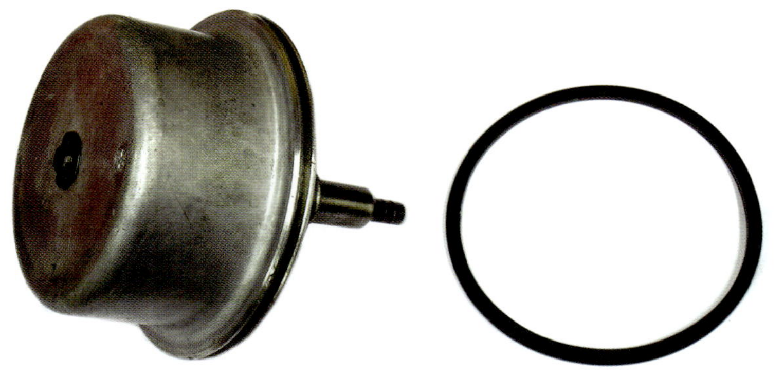

Install a new rubber seal on the low band servo.

5 Install New Piston Seals

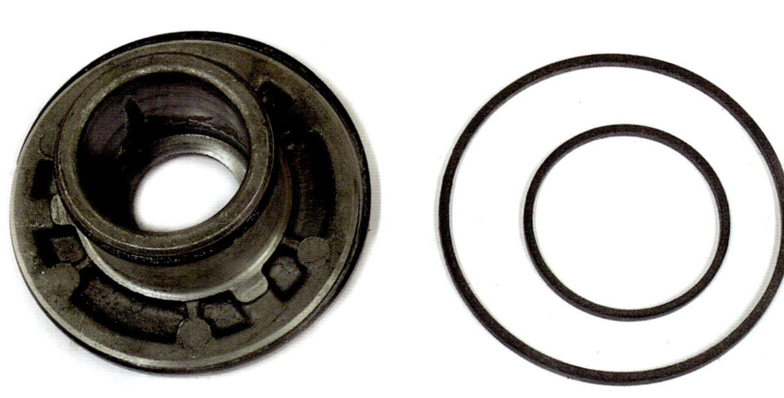

Install new inner and outer seals onto the accumulator piston. Most replacement seals are metal with hooked ends.

Torque Fasteners

6 Install Accumulator Piston Assembly

Lubricate the accumulator piston and slide it inside the band apply piston. Install the entire assembly into the case using a new metal gasket between the cover and the case. Gently tighten all of the retaining bolts evenly, and then torque them to 15 to 18 ft-lbs.

Proper Torque Technique

When tightening any fasteners into an aluminum transmission case, use extreme care because the case may not accept the full torque requirements of the fastener. Older case bolt holes often strip out when bolts are torqued excessively or over tightened.

Install Filter and Valve Body

1 Locate Gaskets

 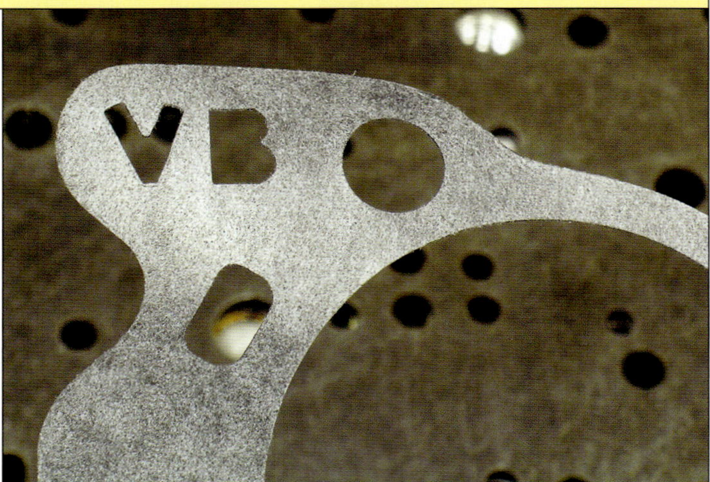

Locate the upper and lower separator plate gaskets. They are marked "C" for the case gasket (left) and "VB" for the valve body gasket (right).

2 Install Check Balls

Install six steel check balls into the case at the locations shown.

TECH TIP

Shift Improver Kit Advice

If you are using an aftermarket shift improver kit, follow the directions in the kit for check ball use and location. Also, make sure you use the valve body gaskets supplied with the kit rather than the ones supplied with the rebuild kit; they may be different.

---

3 Install Gaskets

Install the case gasket, then the separator plate, and then the valve body gasket. Keep the parts together with a small amount of transgel or petroleum jelly.

4 Install Downshift Solenoid Gasket

Install the metal gasket for the downshift solenoid. Some solenoids have their own gasket and do not require an additional one. Make sure that the separator plate and lower gasket do not slip, and that their holes are well aligned, and then tighten the two solenoid retaining bolts. If the solenoid is being used, hook up the wire to the terminal in the case plug.

5 Remove Retaining Clip

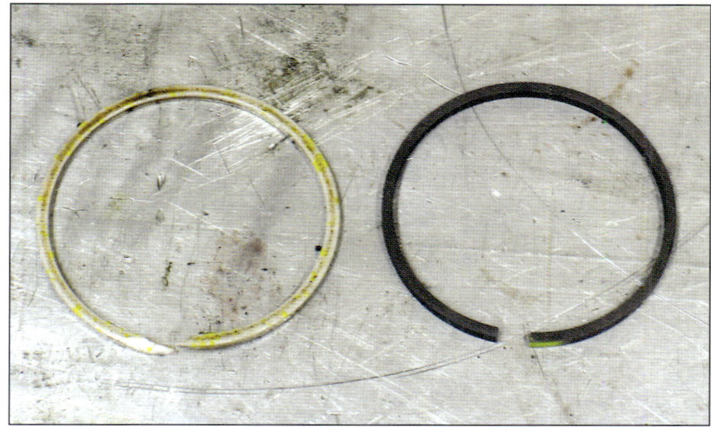

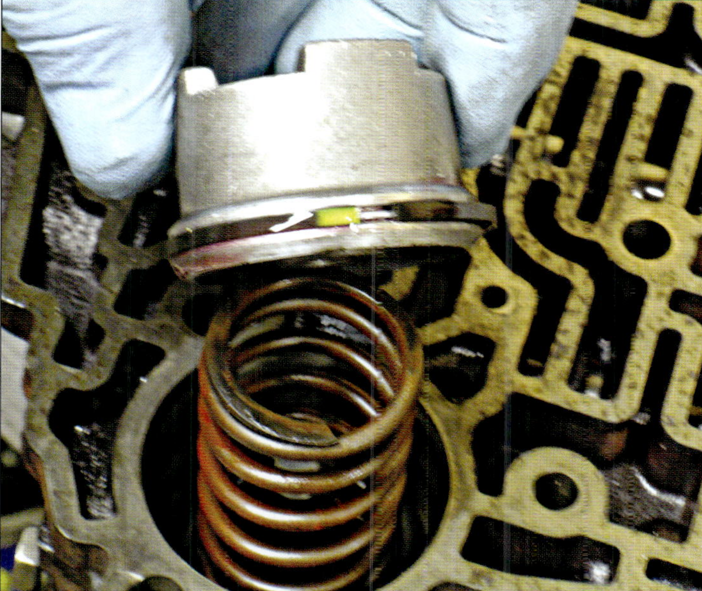

Using a press, large pair of channel locks, or other suitable tool (shown), compress the valve body accumulator and remove the retaining clip. Install a new sealing ring on the accumulator piston, and re-install it into the valve body. Compress the accumulator and spring and re-install the retaining clip.

Accumulator Pistons

Some TH400s use a plastic accumulator piston in the valve body. This is *not* recommended for any level of rebuild and should be replaced with an aluminum piston. Make sure to check any piston used for a tight fit on the pin; a loose fit can cause a leak and negatively impact shift performance.

Professional Mechanic Tip

6 Modify Valve Body

Some aftermarket shift improver kits contain a modified 1-2 shift valve (top left) and plug for the bleed hole in the valve body. Installing this valve and blocking the feed hole (right) does not allow the transmission to up shift into second gear when the selector is in the low-gear position. This basically allows manual control of the 1-2 upshift above the governor speed. Note that this modification allows the transmission to shift to low gear at any *vehicle speed when the shifter is moved to the low position.*

7 Position Governor Filter Screen

Prior to installing the valve body, place a new governor filter screen in the case. The correct location is in the hole closest to the output shaft.

8 Install Valve Body

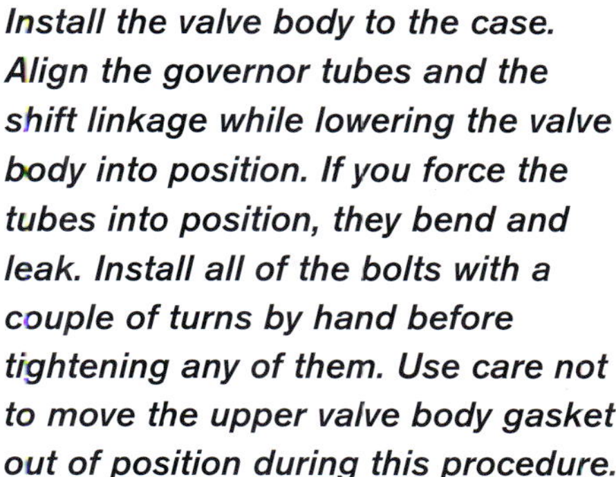

Install the valve body to the case. Align the governor tubes and the shift linkage while lowering the valve body into position. If you force the tubes into position, they bend and leak. Install all of the bolts with a couple of turns by hand before tightening any of them. Use care not to move the upper valve body gasket out of position during this procedure.

9 Install Filter Tube

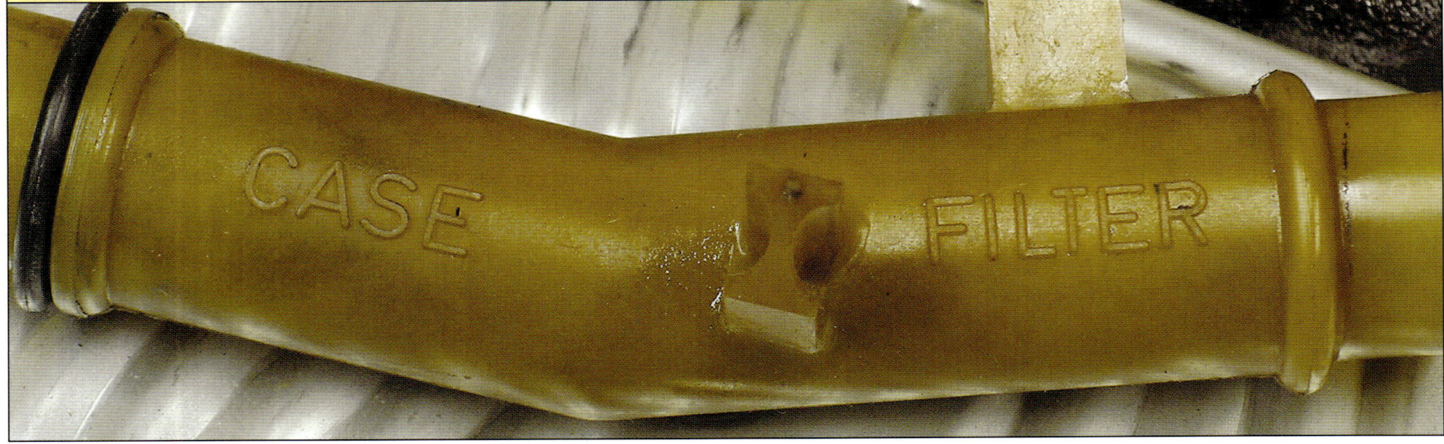

Install the filter tube into the case with a new O-ring. Use a small amount of lubricant to avoid damaging the O-ring during installation. A damaged and leaking O-ring results in poor transmission performance and potential damage after the unit is placed in service. Note that on 1964–1966 models with the one-piece side-mounted filter, the O-ring is placed directly on the filter tube.

10 Install Filter

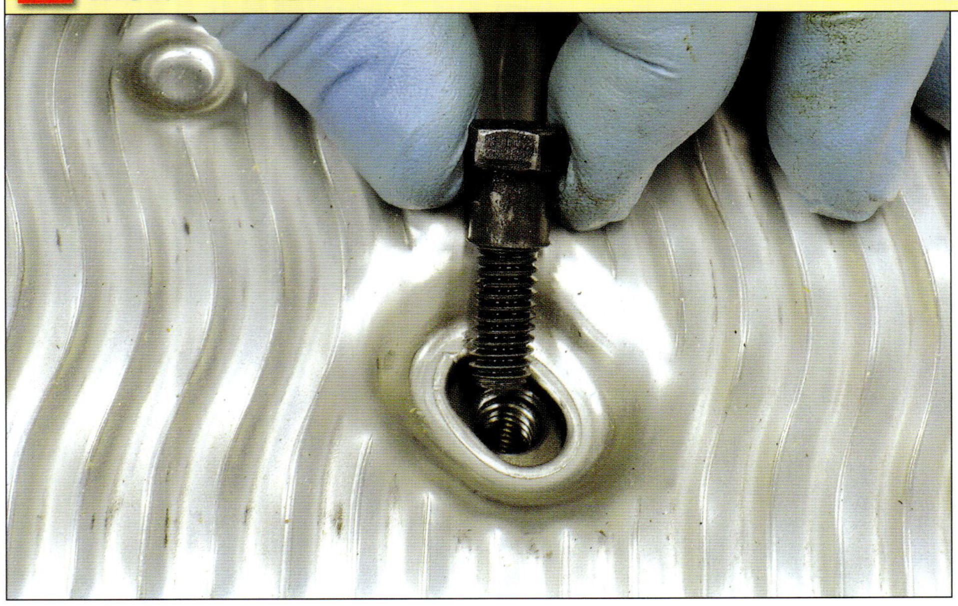

Install the filter over the filter tube and retain with the special bolt. Do not overtighten this bolt; it can be easily broken off in the valve body.

Install Exterior Components

1 Install Modulator Valve

Install the modulator valve into the case. Push the modulator into the case with a new O-ring. Install the hold-down bracket and tighten the single 5/16-inch bolt to hold it in place.

2 Install Governor

Inspect the governor drive gear for wear and replace it if the gear teeth show wear or damage. The gear is replaced by driving out the roll pin and pulling the gear off the governor. The new gear is not drilled for the roll pin, and requires a small hole drilled through it prior to driving the roll pin in place. In most cases it is not necessary to take the governor apart. Clean it thoroughly and make sure all parts move freely and both springs are in place during assembly. Install the governor into the case, meshing it with the drive gear on the output shaft. Install the governor cover, gasket, and four retaining bolts.

3 Install Speedometer Gear Housing

Place a new outer seal onto the speedometer gear housing. Install a new inner lip seal and metal retainer inside the housing. Lubricate the speedometer driver gear and install into the housing. Reference the marks on the housing end and place them toward the retainer. Lubricate the outer seal and push the speedometer gear housing into the case. Install the retainer and tighten the bolt. The retainer ends should mate up squarely with the corresponding locations on the housing.

4 | Install Tail Housing Gasket

Check the engagement of the governor gear and speedometer-driven gears. Place a new tail housing gasket on the case.

5 | Replace Bushings

Using a sharp punch, remove the tail housing seal and bushing. Drive in a new bushing and seal.

Be Gentle with the Tail Housing Seal

Avoid pounding on the seal once it is fully seated; doing so may dislodge the spring inside the lip of the seal. If this happens, make sure to work it back into position before installing the tail housing on the transmission.

6 Install Tail Housing

Install the tail housing and tighten all six bolts securely.

7 Install Oil Pan Gasket

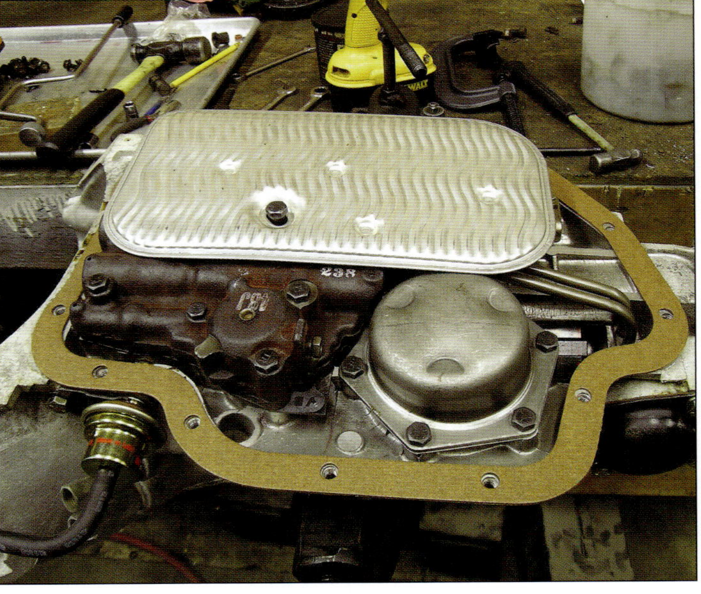

Install a new oil pan gasket. The thick fiber gasket is recommended rather than cork or rubber. Install the oil pan and gently tighten all the oil pan bolts.

Pan Bolt Tightening

Avoid overtightening the transmission pan bolts. No sealers are needed on the pan gasket, but a small amount of RTV used on the pan side of the gasket retains the gasket to the pan if it is removed for any reason after the transmission is placed into service. The fiber gasket can be used many times without requiring replacement.

8 Install Filler Tube Seal

Install a new transmission filler tube seal. I recommend upgrading to the multiple-lip seal shown instead of the O-ring-style tube and seal. Firmly push the filler tube into the multiple-lip seal till fully seated. This type of seal is far superior to the early-style O-ring seals because they simply do not leak.

Our TH400 transmission build is now complete and ready for installation. The torque converter must be installed and fully seated into the oil pump prior to installing it into the vehicle.

Shift Kit TH400

I can't remember building a single transmission that didn't include some modifications to improve shift performance. Stock transmissions have always been set up by the factory to provide smooth shift performance to minimize customer complaints. Any sort of harsh shifting almost always results in a come-back under warrantee. Soft shifts generate a lot of heat and consume engine power. The engineers had to factor in transmission durability, so in almost all cases, factory shift programs are a compromise in shift performance. They actually did a very good job, and most TH400s led a long and healthy life when used within the parameters they were originally designed for.

Keep in mind that not all TH400s were used in heavy-duty service, however. Heavy-duty and high-performance applications put additional stress on the transmission, and a shorter/firmer shift helps to transfer that power without a lot of wasted energy, and minimizes wear on the band and friction linings.

Shift performance can be difficult to evaluate simply by feel, as many items other than the transmission affect shift performance. How well the torque converter is coupled at any particular RPM has a dramatic impact on shift performance felt by the operator of the vehicle. Engine power pro-

duction affects torque converter performance, so it plays a partial role in shift performance. The vehicle's weight, tire diameter, and gear ratio play a role here as well.

Basically, a short shift is typically evaluated as a firm shift. There is always some overlap period during a shift, as one component must release, and another component must apply. Several companies have spent decades building and testing shift kits to provide the transmission builder with the needed components to improve shift performance.

At the builder level, you need to be mostly concerned with shift performance in relation to improving durability of the transmission, at least as a primary goal. It helps to know the fluid path, and power flow through the transmission, but it is not a priority. When you buy and install an aftermarket shift kit, you are paying the company that designed it to do the research and development.

Through the years, many self-proclaimed transmission builders have developed their own modifications to improve shift performance. These modifications typically involve disabling accumulators, increasing separator plate hole sizes, tossing out a few check balls, and shimming up pressure regulator springs. They may also include custom modifications to double feed the direct clutches with the TH400 transmission. The problem

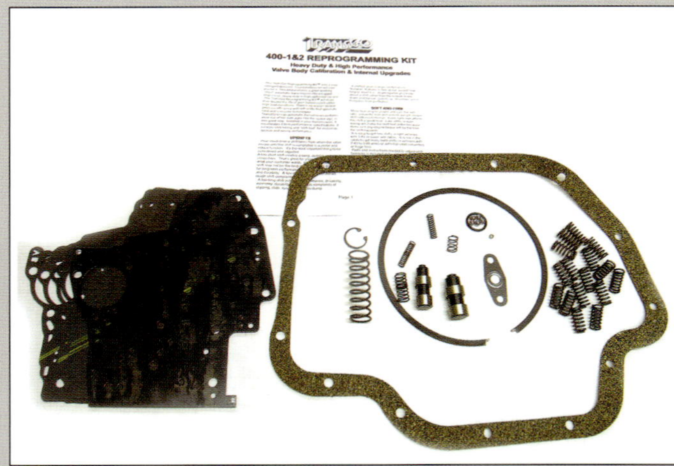

The TransGo Shift Kit comes in a box with an organizer tray to keep all the components separate. They are grouped and bagged accordingly. It also comes with very specific instructions, which also include some additional information for setting up TH400s during transmission overhaul. Since the kit is "adjustable," choose the correct course of action based on the application, and follow the directions to the last detail.

associated with installing your own shift kit, is that you seldom have specific guidance. Some home modifications may work perfectly for a 3,000-pound car, high-stall torque converter, and some tall gearing; and work quite poorly for a heavy tow vehicle, small engine, and stock torque converter. I do not cover any home-brewed shift kits for that reason. This doesn't mean that you can't experiment in that direction if you have the time and resources to do so.

The TH400 is an excellent transmission, and factory shift programs for them typically provide good shift performance for stock or near-stock applications. When you use the TH400 in a performance application, or heavy-duty towing application, improving shift performance also improves the durability of the unit.

TransGo has been making shift kits for transmissions as far back as I can remember. They make several levels for the TH400 unit. The 400 Stage 1 and 2 kit are covered here. It is very easy to install, and can be calibrated for a wide variety of applications. It also incorporates a patented dual-feed system for the direct clutch. This increases holding power of the clutch pack under load and significantly improves clutch pack life in heavy-duty and high-performance applications.

It comes with a spring option for the pressure regulator. The orange spring is dubbed "race only." Here you limit the use of the orange spring to vehicles with excellent power-to-weight ratios that will see some drag strip use. For street vehicles, the stock pressure regulator spring is used. Make sure to re-install any horseshoe shims with the stock PR spring, but remove them if the orange spring is installed.

In all cases leave the valve body accumulator system stock, and only make adjustments in that area if a test drive indicates a slight adjustment is needed. To date this has never happened to me.

Separator plate hole sizes are based on the use of the vehicle. The instructions are very specific as to vehicle usage and separator plate hole sizing. Always stay on the lower range provided, as you can always open up holes later if shorter/firmer shifts in any particular area are desired.

Five check balls are retained with this kit, and both accumulators are left operational. For race-only applications, I have found that the installer can block off the 2-3

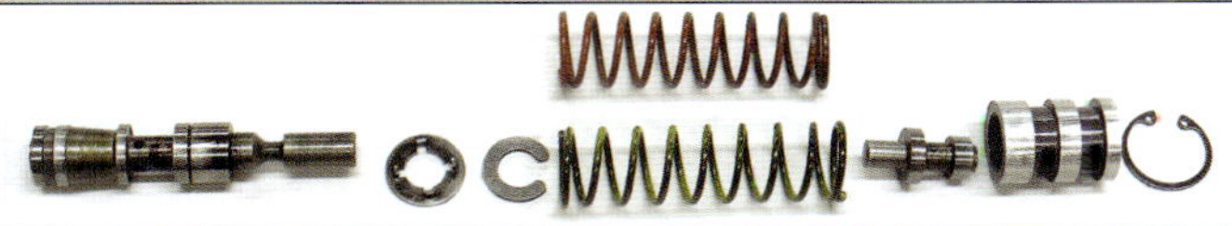

The orange spring is labeled "Race only." It raises line pressure significantly across the board. This decreases shift time and improves the clutch pack and servo clamping force required for racing applications.

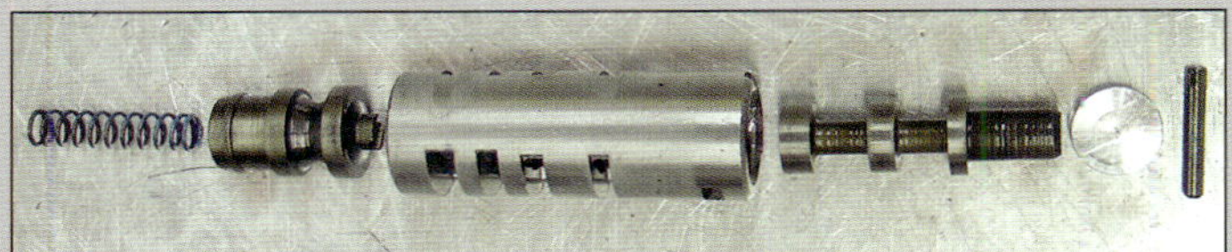

The accumulator system is located in the valve body. Three different arrangements were used over the years of production. The instructions show the arrangement of the accumulator springs required to get the desired results. I always recommend road testing the vehicle before making any modifications to the accumulator system. In many cases, no further modifications are required.

The PR valve and its related components are located in the transmission's oil pump. The pump does not need to be taken apart to install the shift kit. It can be easily accessed with the transmission in the vehicle and the oil pan removed. If the transmission is apart for rebuilding, gently clamp the oil pump in a soft-jawed vise and remove the snap ring that retains the PR assembly. Install the parts in the same order they were removed, including any horseshoe-shaped shims, if re-using the stock PR spring. The shims are not used with the high-pressure orange spring.

The separator plate is a custom part, and significantly different than the stock one you are removing. Drilling the appropriate plate hole sizes is clearly outlined in the instructions with specific guidelines per application. It is always best to start out somewhat conservative. An extra set of gaskets is provided in case they are needed.

The 1-2 accumulator is located inside the servo for the rear band. It should be left functional, and it is a good idea to inspect and replace the seals if needed. Some units used Teflon seals, and they can become loose in the bores causing delayed or "soft" 1-2 upshifts.

Most later-production TH400s use a roller-type intermediate sprag. They are very strong, but don't spread the load as effectively as the 34-element sprag-type clutch used on early units. They are quite effective at holding the load in second gear, but can fail when hit really hard on the 1-2 upshift. This is a good reason to be conservative when setting up the 1-2 shift program.

accumulator with no ill effects to provide a shorter/firmer 2-3 shift.

It is not recommended that you disable the 1-2 accumulator, as it can be incredibly hard on the intermediate sprag assembly. 1-2 shift performance in the TH400 transmission is also affected by which clutch pack arrangement is used during the build. The stock smooth frictions are "waved," and often waved steel is used with that clutch pack. Aftermarket high-performance frictions, and stock waffled frictions are flat. Using flat steels and flat frictions provide the shortest/firmest shift. This is not necessarily the best scenario for the intermediate clutch pack in a race application, as once again, it can be incredibly hard on the

For all levels, five check balls are used in the locations shown. One check ball is omitted just above the intermediate band servo location in the case.

This modification is not part of the TransGo shift kit instructions. The 2-3 accumulator feed hole can be blocked off if a firmer 2-3 shift is desired. This option should only be performed for racing units or where a somewhat "loose" torque converter has been installed. For all other applications, it is best to leave the accumulator functional.

sprag in high-horsepower applications where a tight torque converter is being used.

The intermediate sprag assembly is, and always has been a "sore" spot for these transmissions. The best advice I can provide here is to be conservative with the 1-2 shift as far as separator plate hole sizes are concerned. Using stock smooth/waved frictions is also a good idea for most builds, as this cushions the apply slightly. A single, waved steel plate with flat frictions accomplishes the same thing.

Most builders in recent years actually recommend using one waved steel plate in each clutch pack for the TH400 when it is used in a high-performance application.

Any high-performance TH400 build should also include a smooth direct drum and 34-element sprag

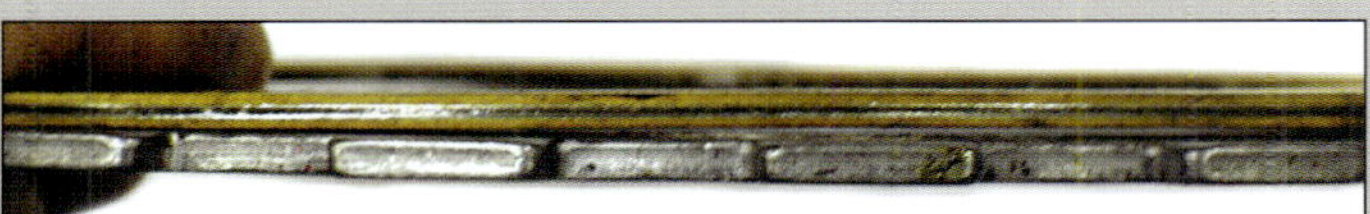

Factory-type smooth intermediate frictions are waved slightly. This acts as a cushion to soften and slow the shift slightly as all of the components of the intermediate clutch pack need to be fully compressed to complete the shift. Factory-type waffled frictions are flat, as are most aftermarket "high-performance" frictions. This fact affects shift performance, and should be factored in when drilling the separator plate hole for the 1-2 shift.

Early TH400s used a smooth direct drum and sprag-type clutch. An aftermarket 34-element clutch is a highly recommended upgrade. 4L80E transmissions came with a smooth drum and the 3-element sprag-type clutch as standard equipment. The drums are available new and can be installed directly into a TH400 without any modification.

assembly, retained with the latest design 4L80E spiral lock ring. Both of these modifications help to improve the durability of the sprag assembly in a high-performance application. There's nothing at all wrong with using the same setup in a stock application, as the sprag assembly

Some units also use a waved steel apply plate in the intermediate clutch pack. This functions in the same manner as the waved smooth frictions, softening or cushioning the 1-2 upshift.

Selecting Shift Firmness

Choosing feed hole size: Be conservative. It's easier to produce a tight even feeling shift with some "class" if the feed hole is not too big.

Hole A: 2nd feed hole size

Range #1: Passenger, police, taxis, van, limo, and light trucks: leave as is or .082

Range #2: Motor homes, heavy trucks, street rods, & huge tires .093 - Comfort at light throttle and shorter at heavy throttle.

Range #3: .110 Hot Rods Starts getting short above 1/4 throttle. Takes careful accumulator adjustment to get comfort.

Range #4: .156 Race/Competition only.

Hole B: 3rd feed hole size

3rd clutch holding power has been tripled, so be conservative with hole size. After a road test you can always make it bigger. Extra gaskets are furnished for changes.

Range #1 Passenger, police, taxis, van, limo, and light trucks: Don't drill 3rd hole.

Range #2 Motor homes, heavy trucks, street rods, & huge tires: Don't drill 3rd.

Range #3 Hot Rods You want it firm: .082

Range #4 .156 Race/Competition only.

Hole C: 3rd Accm feed size
Range #1, Drill feed .125 - .140
Range # 2, 3, 4: Don't enlarge accm feed.

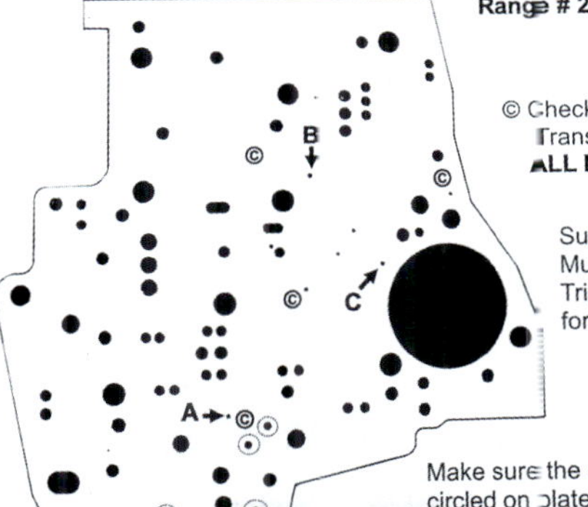

The separator plate hole sizes are detailed in the instructions per specific application. The guidelines should be followed closely, leaning toward the smaller hole sizes. You can always open up the hole sizes later, after road testing the vehicle.

Shift Kit TH400 *continued*

is much stronger than the later-design roller clutch that showed up in the TH400's built after 1971.

The supply of original-equipment smooth drums is limited. The aftermarket 4L80E smooth drum is readily available, and cost effective. It's proven itself as a durable part, and is a direct fit into a TH400 transmission.

The TransGo TH400 shift kit comes with quite a few parts, and detailed directions as to which ones to install for a specific purpose. Two options are to re-use the stock pressure regulator spring, or to install the orange spring. The orange spring is specifically used for "race only" applications.

The accumulator system is also addressed in the instructions, and comes with several spring options and specific guidance as to how to install them. Even though this is an option to fine tune shift performance, I have always left the stock accumulator system in place, and road tested the transmission first. To date I have never had to modify it.

The kit comes with a new steel separator plate. Options for drilling feed holes are outlined in the instructions. It is best, as recommended by the directions, to be conservative with plate hole sizes. It's much easier to make them larger later, and extra gaskets are provided in case they are damaged.

It is also imperative to use the supplied gaskets, as they are different than the ones that come in the rebuild kit. If any other separator plate gaskets are used beside the TransGo–supplied ones, the transmission will not function correctly!

This kit also provides full control of first and second gear. This is done by installing a custom 1-2 shift valve, and driving a plug into the bleed hole in the valve body.

Installing the custom 1-2 shift valve provides full manual control of first and second gear. The transmission no longer upshifts from first to second if the shift lever is left in the "L" position. All automatic upshifts are still retained. Be aware that installing the 1-2 shift valve allows for a shift into first gear at any vehicle speed. If you do not want this function, do not install the custom 1-2 shift valve and leave the bleed hole open.

Over the years I have installed the TransGo shift kits with complete success. I followed the directions very closely, specifically for separator plate hole sizes. The only time I use large plate hole sizes is for full-race applications

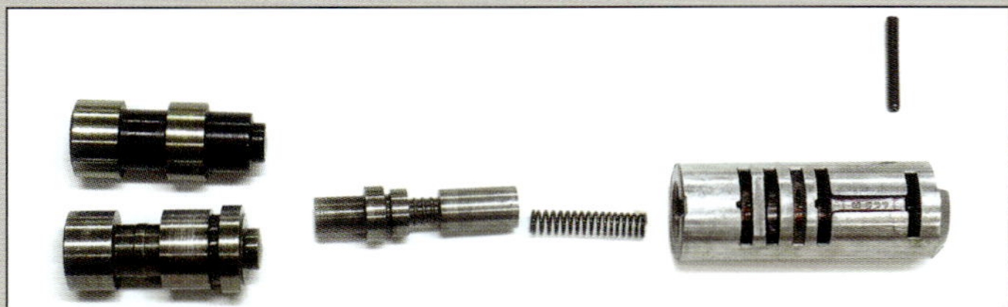

A modified 1-2 shift valve is included with the kit. There are two options, and you should replace the stock valve with the one that matches it in diameter. The modified 1-2 shift valve provides full manual control for all upshift functions above governor speed. This eliminates the factory built-in 1-2 upshift even if the transmission is left in low gear. This is a must for drag racing units that are going to be manually shifted. However, be aware that installing this valve puts the transmission shift into second or first gear anytime the shifter is moved to those positions at high vehicle speeds. This means if you are going 100 mph and move the selector into the L1 position, the transmission shifts immediately into low gear. If you do not want this to be possible, then do not install the new 1-2 shift valve.

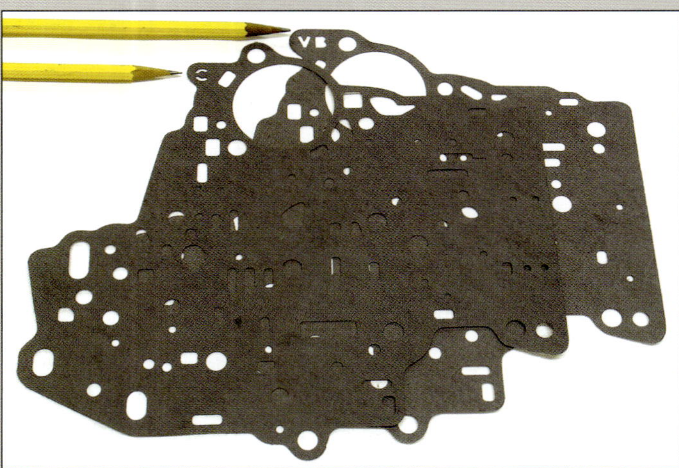

Trans-Go supplies two sets of gaskets with their shift kit. Do not use stock gaskets, as poor shift performance and transmission failure may occur.

with relatively "loose" torque converters. A loose torque converter softens the shift "feel" a bit right from the start. For heavy vehicles with stock or relatively "tight" torque converters, especially towing applications, the minimum range of hole sizes should be used.

The kit can be installed with the transmission in the vehicle. It is best done during transmission rebuilding, as there are additional parts supplied in the kit that are installed inside the transmission.

One of those parts is a heavy-duty snap ring for the case. It is installed after the intermediate clutch backing plate. The heavy-duty snap ring locks into the case with much greater authority than the stock part. It's a very good part, and provides additional integrity and support to the intermediate clutch pack.

A set of stronger springs are supplied for the direct clutch. If the transmission is apart for rebuilding, install the stronger springs. They are designed to prevent direct clutch apply at high RPM, and to provide a quicker 3-2 kickdown action.

I use and recommend an adjustable vacuum modulator. Install the pink spring from the kit. It can be removed after road testing if later or shorter part throttle shifts are desired. The adjustable modulators provide some adjustment for light part-throttle upshifts. By using the pink spring on the modulator valve, and the adjustable modulator, it only takes a few quick test drives and minor adjustments to get the results you are looking for.

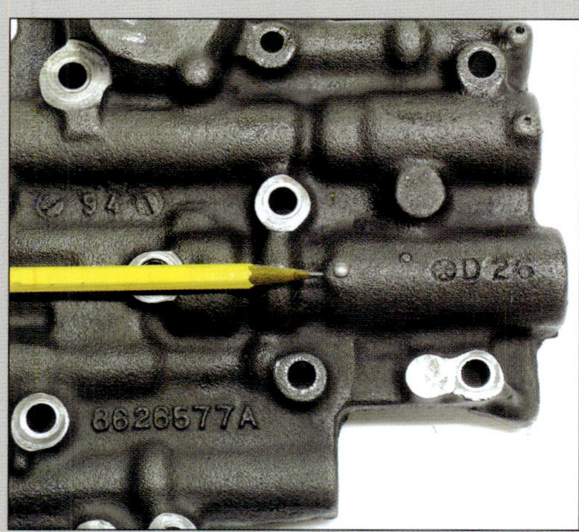

When the modified 1-2 shift valve from the kit is installed, you must drive the small plug from the kit into the vent hole in the valve body at the location shown.

Another excellent feature of this kit is the inclusion of a heavy-duty snap ring for the case. The new ring is significantly stronger than the stock snap ring, and helps to keep the case lugs from getting pounded out above the snap ring by the application of the intermediate clutch.

If the transmission is being rebuilt, replace the stock direct drum piston return springs with the ones supplied in the kit. They are designed to prevent centrifugal apply of the direct clutch at high RPM.

Install the small pink spring on the stock modulator valve. After test driving the vehicle, it can be removed if later/shorter/firmer light throttle shifts are desired. It is also a good idea to install a new adjustable modulator when the shift kit is installed. This allows you to custom tune light part-throttle upshifts.

TH400 TRANSBRAKE

Transmission brakes (transbrakes) have been available for the TH400 for decades. The concept of using a transbrake is to produce the fastest possible acceleration from a standing start. Transbrakes are used specifically for drag racing. A transbrake is activated by a driver-controlled switch that connects to a solenoid mounted on the transmission. When engaged, the transbrake applies forward/low gear and reverse at the same time. Working against each other, the vehicle is unable to move, and is effectively locked in position on the starting line. When the brake is released, reverse is released. This allows the vehicle to move forward in normal fashion.

Now, imagine what would happen if the throttle were applied while the transbrake was engaged, and the RPM levels were up against the full stall speed capabilities of the torque converter.

With the throttle on the floorboard, and the engine up against the converter, releasing reverse instantly allows the vehicle to apply full power to the tires, for the best launch possible with the combination of parts being used.

This is much more effective than trying to use the vehicle's brakes to hold the car on the starting line while simultaneously increasing the throttle and bringing the engine up on the converter instead. This practice is only marginally effective because the engine power through the transmission eventually overcomes the brakes and pushes the car forward. This is not a good scenario if you are staged on the starting line at the drag strip and waiting for the lights to come down.

With well-chosen parts, a transbrake can be used to achieve the best possible vehicle launch in the minimum amount of time. When the engine is already at full throttle, and up near peak torque, there is much less lag time than being up against the converter and on the brakes at a lower RPM. When braking, the converter has to flash to its full stall

The heart of the transbrake is the solenoid. It mounts to the outside of the TH400, replacing the vacuum modulator. It controls a special modulator valve supplied with the TCI kit. Being electrically controlled by the driver, it provides lightning-fast launches at the dragstrip.

The transbrake to be installed is from TCI Automotive, and carries part number 221500. TCI is a supplier of aftermarket torque converters and other transmission components. Their transbrake kit is complete, and comes with detailed picture instructions.

speed and this takes some time. With a transbrake, you're already sitting on the starting line at full power waiting to release the brake and apply full power to the tires.

How does a transbrake work? It's actually a very simple device, using an electric solenoid to place the transmission in low gear and reverse at the same time. Since an electrical solenoid is used, the release time for the brake is nearly instantaneous.

Many racers also employ other electronics in conjunction with the use of a transbrake. They may use a two-step (or RPM limiting) device to keep the engine from exceeding a predetermined RPM while the brake is engaged and the engine is at full throttle. They may also include a delay, which predetermines the exact time from when the transbrake button is released, to when the transbrake releases. At the exact moment the brake releases, so does the two-step. This practice is used most often in Super Pro Drag racing, and allows some deadly consistent reaction times. This practice is so deadly consistent that most race tracks have rules limiting the use of delay boxes

and transbrakes to the higher-level classes. Some tracks even have what's called a "box" (or electronic aid-specific) racing class specifically for these cars.

It is important to keep in mind that other vehicle modifications may be required in addition to the transbrake. Since you are basically applying full engine power to the tires at the loaded stall speed of the torque converter, traction may become a problem. It's quite common to use larger tires, improved tread compounds, and suspension modifications in conjunction with a transbrake. Parts breakage also becomes a real possibility, since the transbrake is going to shock the tires with near-maximum engine power right on the starting line. From my experience, any weak links in the vehicle's drivetrain quickly show themselves on the first track outing!

Installation Procedure

Installing a transbrake requires several internal modifications. It is best to install the transbrake while the transmission is being rebuilt.

The TCI transbrake installation kit contents are show here. It contains a new modified valve body, solenoid, separator plate, separator plate gaskets, springs for the direct apply piston, solenoid brake valve w/spring, pressure regulator spring, and one 1/4-inch steel check ball.

1 Drill Bleed-Off Hole

A bleed-off hole is required in the direct drum under the apply piston. Some drums have a check ball located in the drum, but the bleed-off hole is still required. Using a 1/16-inch (.0625-inch) drill bit, drill a hole from inside to outside under the apply piston. Deburr the hole and blow out any metal chips from the drilling process.

2 Open Partition

With a 1/8-inch round file, open up the partition in the picture. Do not scratch or nick the bore where the brake valve is located. Remove at least 50 percent of the partition and clean the case thoroughly to remove all the metal shavings.

3 Remove Center Seal

Remove the center seal on the direct drum. No seal is used at the location shown. Install new inner seals and outer seals on the direct drum apply piston.

4 Install Springs

Install 16 new gold-color springs on the direct drum apply piston, and install the retainer and snap ring. Set the clutch pack clearance to .050 to .080 inch.

5 | Remove Band

Remove and discard the intermediate band and the apply piston in the case. They are not used with a transbrake application.

6 | Remove Seals

Remove both seals from the accumulator piston located inside the rear servo assembly. Some seals are metal here. Some pistons use solid Teflon seals. Carefully cut them with a pocket knife and pull them out of the grooves. No seals are installed on the piston during assembly.

7 | Install Check Ball

Drive the 1/4-inch steel check ball provided into the oil passage at the bottom of the accumulator bore in the case.

8 | Smooth-Out Servo Cover

Make sure that the rear servo cover is flat. Attach a piece of fine grit sandpaper to a flat surface and gently work the cover in a figure-8 motion until it is flat and smooth.

9 Replace Modulator Valve

Remove the stock modulator valve from the case and replace it with the TCI brake valve and spring.

10 Install Solenoid

Install the TCI solenoid over the new brake valve. Use the stock modulator retainer and bolt to hold it in place.

11 Remove Center Support Ring

Remove the second ring from the center support assembly. Shown here is a Teflon ring. Carefully cut the ring with a pocket knife and pull it out of the groove. Hooked iron rings may be used here as well. If so, just unhook the ends and remove the ring by spreading it out and lifting it off the center support. Remove the upper ring first; then put it back in place.

12 Remove Snap Ring

Gently clamp the oil pump in a soft-jawed vise or between two blocks of wood. With a pair of snap ring pliers, remove the snap ring over the PR valve assembly. It may help to push down on the assembly with a flat tip screwdriver while removing the snap ring.

13 Replace Pressure Regulating Valve Assembly

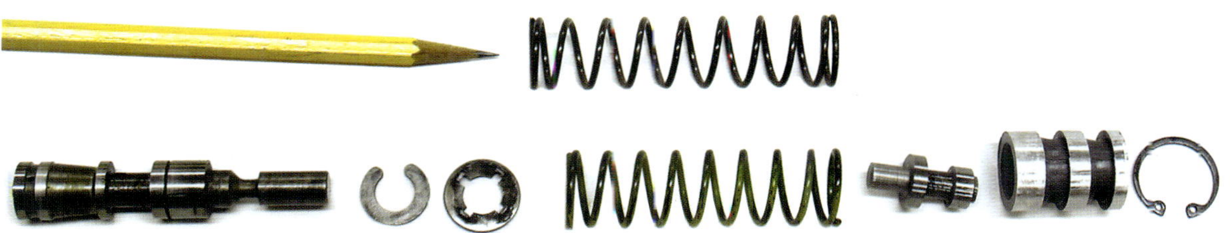

Remove the pressure regulating valve assembly from the oil pump. Replace the stock pressure regulating spring with the one from the TCI TransBrake kit. The assembly may include horseshoe shims below the spring retainer. (Refer to step 6 on page 94 for more information on converter feed-hole modifications.)

14 Use Supplied Gaskets

The TCI-supplied valve body gaskets must be used with the TransBrake. They are clearly marked "Case" and "VB," and must be installed in the correct locations.

TECH TIP — Using Reverse

The TCI 221500 valve body is equipped with a special reverse apply circuit. In order to apply reverse follow these steps:

- Put shifter in neutral.
- Push transbrake button to back up.
- **Do not** use the reverse shifter position.

Even though it is possible to engage reverse by using the reverse shifter position, clutch and band damage may occur.

15 Install Separator Plate

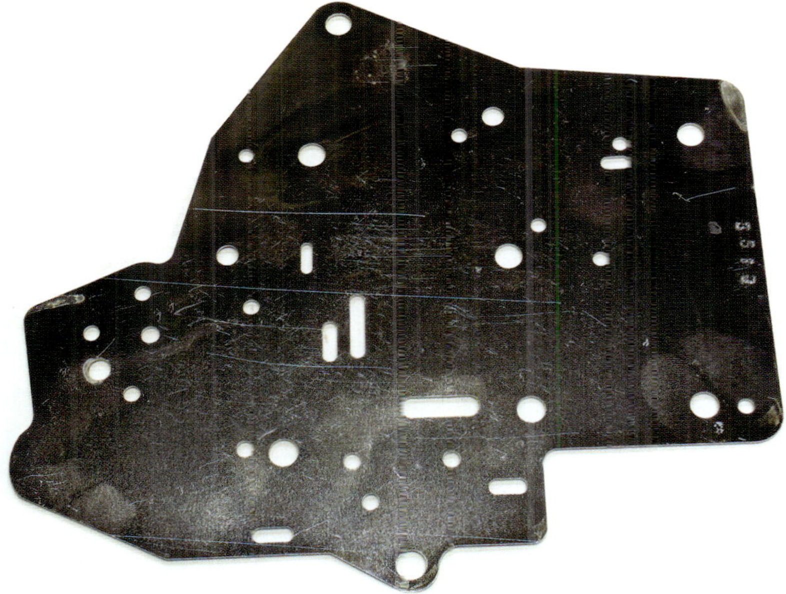

The separator plate from the TransBrake kit must also be used. No additional drilling or other modifications to it are required.

TECH TIP — Warning

With the TCI 221500, do not shut down the motor while in second gear with engine RPM above 6,000. Repeatedly doing this may cause severe damage to transmission and bodily harm A TCI 980000 Aluminum Trans-Shield is highly recommended for this application.

TRANSMISSION INSTALLATION

As with most things associated with this hobby, it takes much longer to install something than it does to remove it. For transmission installation, this is where you see the biggest benefits of using a lift and transmission jack. It can support the weight for hours and lift the unit slowly into place while you get everything lined up and other items out of your way.

The first step in this process is to install the torque converter into the transmission. Slowly pour at least 1 quart of clean ATF into the torque converter. Lubricate the bearing surface of the hub and slide the torque converter onto the transmission. At the same time spin the torque converter to help align the input shaft and stator splines. This may require considerable effort. Make sure that the notches in the hub engage with the transmission oil pump.

Our transmission is in place, strapped down, and ready to safely install in the vehicle.

1 Pour Transmission Fluid

Pour at least one quart of automatic transmission fluid into the torque converter before installing it into the transmission.

2 Install Converter

The torque converter must be installed and fully seated into the transmission before it can be installed in the vehicle. This may take several minutes, as you need to align the input shaft, stator shaft, and oil pump drive. Turn the converter while pushing it rearward at the same time.

If the torque converter is new, and a new bushing was installed in the oil pump, the torque converter is going to be a relatively tight fit, and can be considerably difficult to install. You may have to be patient during this procedure. Several minutes of spinning the converter and pushing it back into the transmission may be required to get it to fully engage. Once fully seated, there is very little room between the back of the converter and the oil pump.

3 Note Space between Converter and Oil Pump

When fully seated in the transmission, there should be very little room between the back of the torque converter and the oil pump.

It may help to place the transmission on the tail shaft and lean it back against a wall when installing the torque converter. This allows the weight of the converter to aid in getting it fully seated onto the input shaft, stator shaft, and oil pump drive.

4 | Use Caution

The torque converter can be very difficult to install into a freshly rebuilt transmission. Gravity can be our friend here, and it may be necessary to stand the transmission on end. Here the bellhousing has been strapped to one of the lift posts to install the torque converter.

Once you have installed the torque converter, spin it several times to make sure there are no tight spots or binding. Do not attempt to install the transmission unless the converter is fully seated. Any attempt to pull the transmission up to the engine with the bellhousing bolts can cause oil pump damage.

Place the transmission on the jack and secure it with a chain or strap. Use care here to ensure the torque converter doesn't fall out during this procedure. Adjust the transmission jack so the transmission is angled back slightly while it is raised into position. (If you are working by yourself, you may want to temporarily install one bolt and a piece of flat bar (bent in slightly) to hold the torque converter in the transmission.)

Safety Step

5 | Use Strap to Hold Converter

A retaining strap is used to keep the torque converter from falling out of the transmission until you are ready for the installation. A piece of aluminum flat bar with a 3/8-inch hole in it works well for this purpose.

Determine if the filler tube needs to go in before lifting the transmission in place, or if there is room to install it afterward. With many modern vehicles, the curve of the bellhousing and proximity to the firewall does not allow the tube to be installed after the transmission is in place and bolted to the engine.

Another big problem can occur when upgrading the torque converter. If the stock torque converter is being replaced, the new converter should be test fitted to the back of the crankshaft to make sure the hub seats fully. This little detail is often overlooked until the transmission is bolted in place and it's time to install and tighten the torque converter bolts. Some flexplates do not have the needed bolt pattern for the new converter. Some high-performance converters are also set up for use with larger-diameter bolts than factory converters. It's better to find out that the flexplate holes need to be opened up before you are ready to push them through and tighten them up!

6 Test Fit Torque Converter

If you are installing a new torque converter, make sure to measure the front hub on the torque converter and recess on the back of the crankshaft. A test fit between the two components is a good idea, before installing the transmission. The converter should fit tightly into the crankshaft, but still fully seat and turn freely. This step is extremely important because the relationship between the converter and crankshaft maintain perfect alignment between the parts.

7 Note Bolt Hole Sizes

Check the size of the holes in the flex plate before installing the transmission. Some aftermarket converters require larger bolts and the flex plate needs to be drilled to accept them. Find this out before installing the transmission and bolting it in place!

To help guide the transmission onto the engine, make a couple of guide studs. With a hacksaw, cut the heads off a couple of 3/8-inch bolts and grind the cut end to a point. Slot them with a hacksaw or cutting wheel so they can be turned with a flat-blade screwdriver. Install them just above the dowel pins on each side of the engine block. As the transmission is raised into position, slide it over the dowel pins. Use the adjusters on the transmission jack to keep the transmission well aligned with the engine. Slide the transmission forward on the alignment studs and onto the dowel pins. Then install at least one bolt on each side of the bellhousing.

8 Use Guide Studs

A couple of guide studs can be fabricated from a 3/8-inch bolt. Cut off the end and slot it to accept a flat-tip screwdriver for easy removal after the transmission is in place.

Slowly tighten the bolts evenly and pull the transmission until it is fully seated against the engine. Reach in and spin the torque converter during this procedure to make sure it is not binding in the hub of the engine's crankshaft. If the torque converter locks up at any point during this procedure, stop immediately! The torque converter may not be fully seated into the oil pump. If the bellhousing bolts are tightened with the torque converter incorrectly engaged and not fully seated, oil pump damage will result.

9 Install Bellhousing Bolts

The upper bellhousing bolts can be difficult to access. A 6-point universal 9/16-inch socket can be used to install them. To help support the socket, wrap it with a piece of electrical tape.

After the bellhousing is pulled up against the engine, with at least one bolt on each side, remove the alignment studs and install and tighten the remaining bellhousing bolts. The upper right-hand bolt (at the two o'clock position) passes through the strap on the transmission oil filler tube.

10 Seat Filler Tube

Make sure the filler tube is fully seated into the transmission and a new O-ring or multiple-lip seal has been installed. The strap on the filler tube is held stationary by the bellhousing bolt.

Raise the transmission high enough to facilitate installing the rear crossmember. Some units lack sufficient clearance to install the crossmember with the rear transmission mount in place. It may have to be installed with the transmission slightly elevated and the crossmember in place, and then lowered onto the crossmember.

11 Install Crossmember

After the transmission is bolted to the bellhousing, raise the transmission just far enough to work the crossmember under it and align/install the bolts.

Once lowered onto the crossmember, install and tighten the rear transmission mount bolts. The transmission jack can be removed at this point.

Before continuing with the installation, check the torque converter again to make sure it turns freely, and then install and tighten the torque-converter-to-flexplate bolts. The engine needs to be turned with a large socket on the front of the crankshaft, or with a large flat-tip screwdriver between the bellhousing and flexplate teeth. It is best to install all of the bolts before tightening them to the final torque. This guarantees that the next hole is well aligned to start the bolt as the engine is turned.

12 Install Crossmember Bolts

Install all of the bolts that go through the crossmember to the frame, but do not tighten them. This allows some latitude to align the crossmember-to-transmission-mount bolts. Once all of the fasteners are installed, tighten the crossmember bolts first, and then the bolts to the transmission rear mount.

13 Install Torque Converter Bolts

Spin the torque converter to make sure it is not binding. Install the bolts one at a time but do not fully tighten them until all three bolts have been installed. The engine needs to be rotated to accomplish this.

Proper Converter Clearance

It's also pretty common to have to shim up a new torque converter for pump clearance. This is best done with hardened washers of a consistent thickness. With the transmission in place and tightened against the engine, push the torque converter all the way into the transmission. Observe the gap between the bolt pads and the flexplate. This clearance should be about 1/8 inch. If not, shim it up as needed using hardened washers. This is a critical step. If the converter is pulled forward too far out of the transmission when tightening the bolts, the hub can be disengaged from the oil pump. This is not a good scenario and immediately results in pump damage or lack of transmission function.

Check the distance between the torque converter and flex plate/flywheel with the torque converter fully seated in the oil pump. Some installations require shimming. The clearance here should be no less than 1/8 inch, and no more than 3/8 inch. Use grade-8 washers as shimming material if the clearance is excessive. If the torque converter is too far out of the pump, the inner pump drive gear can be damaged. There must be some clearance here, or the oil pump and/or engine thrust bearing can be damaged. This is a critical adjustment and it's common to have to shim aftermarket torque converters.

If you find that there isn't enough clearance, or the torque converter is tight against the flexplate, stop immediately! Either the torque converter hub has insufficient clearance inside the crankshaft, or it is not fully or correctly seated into the inner oil pump gear. In any case, the torque converter should spin freely when the transmission is bolted to the engine. If it is bound up or tight at this point, the problem must be corrected before continuing with the installation.

In almost all cases, when an aftermarket high-performance torque converter is being installed, there is plenty of clearance and some shimming of the converter is required. Make sure to use grade-8 fine-thread bolts if new hardware is required to bolt the torque converter in place. The fine threads have greater holding power than coarse threads, and hardened bolts can be torqued to specification so they don't come loose. Soft bolts do not hold sufficient torque and should never be used to hold a torque converter to the flexplate.

14 Use Proper Hardware

Some aftermarket torque converters require custom hardware. Make sure to use grade-8 material here. Soft nuts and bolts work loose and can cause major problems.

Install the transmission cooling lines and tighten them securely. The fittings on the transmission should be held stationary with a second wrench to keep them from moving while the lines are tightened.

15 Tighten Cooling Lines

Make sure to hold the transmission fittings to keep them from turning when tightening the cooling lines.

Install and tighten the speedometer cable and any wires at the plug terminal. The TH400s use an electric downshift solenoid, which requires a wire to be attached to the terminal on the passenger's side of the transmission case.

Install the driveshaft slip yoke into the transmission. Test fit for driveshaft length. With the driveshaft fully seated, the rear universal joint should have approximately 3/4 inch of distance between the joint ends and the yoke. This ensures sufficient clearance so the slip yoke doesn't bottom out in the transmission while still providing plenty of engagement with the splines

16 Check Clearance

With the driveshaft fully seated into the transmission, there should be approximately 3/4-inch clearance between the U-joint ends and the yoke.

Driveshaft Dimension Advice

If a new driveshaft is being made for the vehicle, make sure that the vehicle's weight is on the tires when taking this measurement. The shop making your driveshaft is going to want to know the center-to-center distance between the universal joint ends in the slip yoke and differential yoke. Make sure the yokes are well aligned and the slip yoke is pulled back 3/4 inch from bottomed out in the transmission when taking this measurement.

If a new driveshaft is going to be made for the application, fully insert the slip yoke into the transmission, and pull it back approximately 3/4 inch. Measure the center-to-center distance between the front U-joint ends and differential yoke ends. Provide this measurement to the shop building your driveshaft.

17 Add Transmission Fluid

Add 4 quarts of ATF to the transmission and start the engine. Let the engine idle and continue to add fluid until it is on the lower range of the scale on the dipstick. Do not overfill the unit. With the vehicle effectively supported and the emergency brake locked, place the transmission in forward and reverse, then back into park. Re-check the fluid level, and then add as needed. Do not attempt to road test the vehicle until the transmission fluid level is within the full range on the dipstick.

Install the torque converter dust cover and engine starter if it was removed. Reconnect the battery cables. Make sure that the vehicle is well supported on a lift or jack stands, and that the drive tires are not touching the ground. Lock the parking brake.

Add 4 quarts of ATF to the transmission and start the engine. These first few quarts of transmission fluid are picked up and pumped to the torque converter almost immediately. Continue to add transmission fluid with the engine at idle speed until the fluid level registers on the dipstick.

Once at the bottom of the range, apply the brakes and place the transmission in drive (D), then neutral (N), then reverse (R), and back to park (P). Check the fluid level again, and add as needed. Leave the level about 1 pint low to allow for expansion of the fluid when the transmission heats up.

TROUBLESHOOTING GUIDE

Troubleshooting is the step-by-step procedure where you can logically diagnose a problem. This involves a process of elimination, usually starting from the simple, easy-to-check items, and moving on to more difficult items in a logical sequence.

For example, if your transmission doesn't move the vehicle at all when placed into service, you'd check the transmission fluid level before pulling the transmission out to check the oil pump or torque converter.

The TH400 is a very rugged and durable unit, and fairly simple in construction and operation as far as automatic transmissions are concerned. If you have correctly assembled the unit, any issues with it not working should be relatively easy to diagnose and repair.

If you look at the list below, it quickly becomes evident that many problems can share a common cause, such as the transmission being low on fluid, or the manual shaft not moving the manual valve correctly.

Another common problem that can affect TH400 performance at every level is a leak between the case and the oil filter tube O-ring. Some builders even install a second O-ring on the tube when it is pushed into the case, as a backup. A cracked filter tube may be harder to detect by visual inspection, but it is another problem that causes the transmission to not work properly, if at all. The pump picks up considerable air as well as oil from the pan. This usually "foams" the oil, and the transmission cannot reach or maintain correct operating pressures to function properly.

The most difficult problems to diagnose involve the valve body (control valve assembly). In almost all cases, the valve body must be removed to correct any issues, and this should be done after all other possibilities have been eliminated.

Small pieces of dirt or debris are the enemy of the automatic transmission. In many cases involving poor performance, sticking valves, and/or erratic shifting, you find some sort of contamination has gotten into the control valve assembly and restricted proper fluid flow.

In any case, when troubleshooting the TH400, as with most other mechanical devices, start with the simple and accessible things first, then move on to those more difficult to inspect/repair items.

The following is a list of symptoms and their likely causes.

Vehicle Slips in Park

1.   *Parking pawl guide damaged or installed incorrectly*
2.   *Parking pawl broken, cracked, or worn smooth*
3.   *Manual linkage not properly adjusted*

No Drive in any Range

1.   *Manual control linkage out of adjustment or manual valve pin not attached to linkage*
2.   *Low oil level*

---

3.   Low oil pressure, blocked filter, leaking at filter tube O-ring, pressure regulator valve stuck or damaged, damaged pump drive gear or torque converter hub
4.   Defective torque converter

No Drive in "D"

1.   Manual linkage
2.   Low oil level
3.   Low oil pressure, blocked filter, leaking at filter tube, and/or O-ring in case
4.   Forward clutch pack not applying, air check circuit
5.   Sprag or roller clutch assembled incorrectly or not holding

No 1-2 Shift

1.   Governor valve stuck, governor oil tubes leaking
2.   Governor gear damaged
3.   1-2 shift valve assembly stuck closed or incorrect valve body gaskets
4.   Intermediate clutch incorrectly assembled, air check circuit
5.   Intermediate sprag broken or installed incorrectly

1-2 Shift at Full Throttle Only

1.   Detent switch defective or energized
2.   Detent solenoid gasket leaking or bolts not tightened
3.   No vacuum to modulator or stuck/sticking modulator or modulator valve

No Reverse

1.   Low oil level
2.   Check manual control linkage
3.   Defective vacuum modulator
4.   Filter plugged, restricted, or leaking at filter tube or O-ring in case
5.   Damaged rear servo or seal missing
6.   Reverse band damaged, apply pin not engaged, or anchor pins not engaged with band
7.   Damaged direct clutch or leaking check ball in direct clutch piston
8.   Forward clutch not releasing
9.   Low/reverse check ball missing from case

Slips in all Ranges and on Starts

1.   Low oil level
2.   High oil level and oil "foamed"
3.   Plugged filter
4.   Air leak at filter tube or filter tube O-ring in case
5.   Modulator valve sticking
6.   Oil pump damaged
7.   Pressure regulator or boost valve stuck or sticking
8.   Snap ring on pressure regulator valve dislodged from groove in pump
9.   Forward and direct clutch pack clutches burned, damaged, or not assembled correctly
10. Valve body gaskets misaligned or wrong gaskets

Slips on 1-2 Shift

1.   Low or incorrect oil level
2.   Sticking or defective vacuum modulator

3.   Pressure regulator valve defective
4.   Out of position or incorrect valve body gasket(s)
5.   Intermediate clutch piston seal damaged, reversed, or missing
6.   Damaged intermediate clutch plates
7.   Accumulator seal damaged or missing
8.   Damaged intermediate sprag or inner or outer sprag race

Slips on 2-3 Shift

1.   Items 1-4 under "Slips on 1-2 Shift"
2.   Direct clutch plates burnt
3.   Damaged or missing oil seal rings on direct clutch piston
4.   Damaged or mission sealing rings on center support assembly
5.   Damaged or missing accumulator seal

No Engine Braking in Second

1.   Front servo or accumulator sealing rings damaged or missing
2.   Front band broken or lining burnt or damaged
3.   Front band pin not engaged on anchor pin and/or servo pin

No Engine Braking In Low Range

1.   Low/reverse check ball missing from control valve assembly
2.   Rear servo damaged or missing oil seal ring, damaged bore, or damaged piston
3.   Rear band broken, burnt lining, or not engaged on anchor or servo pins

No Part-Throttle Downshifts

1.   Vacuum modulator defective
2.   Modulator valve stuck or sticking
3.   Control valve assembly has stuck 3-2 valve or broken spring

No Detent Downshifts

1.   Detent switch needs fuse, connections tightened, or out of adjustment
2.   Detent solenoid defective
3.   Detent valve in control valve assembly malfunctioning

Low- or High-Shift Points

1.   Low or high oil level
2.   Vacuum modulator assembly malfunctioning, vacuum line connection loose, disconnected or plugged, or wrong vacuum source to vacuum modulator (requires full manifold vacuum)
3.   Detent solenoid stuck open or loose
4.   Control valve assembly malfunctioning, check detent, 3-2 and 1-2 shift valves and springs
5.   Separator plate gasket incorrect or out of position

Noisy Transmission

1.   Noise from pump caused by low or high oil level
2.   Cavitation from plugged filter, leak at filter tube or O-ring in case
3.   Water in oil
4.   Damaged pump or pump gears
5.   Worn or damaged planetary gears
6.   Burnt or worn clutch plates, or burnt or worn clutch lining on bands

SOURCE GUIDE

**Automotive Transmission
Service Group**
18635 SW 107th Avenue
Cutler Bay, FL 33157
305-670-4161
www.atsgmiami.com

Continental Torque Converters
730 Centinela Avenue
Inglewood, CA 90302
310-674-1072
www.ctconverters.com

**Heli-Coil
Emhart Teknologies**
50 Shelton Technology Center
Shelton, CT 06484
877-364-2781
www.emhart.com/brands/heli-coil

TCI Automotive
151 Industrial Drive
Ashland, MS 38603
888-776-9824/662-224-8972
www.tciauto.com

Transtar Industries
7350 Young Drive
Cleveland, OH 44146
800-359-3339
www.transtar1.com